Tobias Johannes Puschmann

Histologische und mikrobiologische Untersuchung von punktionsbedingten Hautstanzen, in Abhängigkeit von Kanülenarten und Desinfektions- und Vorbereitungsmethoden der Injektionsstelle beim Pferd

Cuvillier Verlag Göttingen
Internationaler wissenschaftlicher Fachverlag

Bibliografische Information der Deutschen Nationalbibliothek

Die Deutsche Nationalbibliothek verzeichnet diese Publikation in der Deutschen Nationalbibliografie; detaillierte bibliografische Daten sind im Internet über http://dnb.d-nb.de abrufbar.

1. Aufl. - Göttingen: Cuvillier, 2016

Zugl.: Hannover (TiHo), Univ., Diss., 2016

Nonnenstieg 8, 37075 Göttingen

Telefon: 0551-54724-0

Telefax: 0551-54724-21

www.cuvillier.de

1. Auflage, 2016

Gedruckt auf umweltfreundlichem, säurefreiem Papier aus nachhaltiger Forstwirtschaft.

ISBN 978-3-7369-9408-9

eISBN 978-3-7369-8408-0

Tierärztliche Hochschule Hannover

Histologische und mikrobiologische Untersuchung von punktionsbedingten Hautstanzen, in Abhängigkeit von Kanülenarten und Desinfektions- und Vorbereitungsmethoden der Injektionsstelle beim Pferd

INAUGURAL – DISSERTATION
Zur Erlangung des Grades
eines Doktors der Veterinärmedizin
- Doctor medicinae veterinariae -
(Dr. med. vet.)

vorgelegt von
Tobias Johannes Puschmann
Filderstadt

Hannover 2016

Wissenschaftliche Betreuung: Prof. Dr. Bernhard Ohnesorge
Klinik für Pferde
Tierärztliche Hochschule Hannover

Prof. Dr. Christiane Pfarrer
Institut für Anatomie
Tierärztliche Hochschule Hannover

1. Gutachter: Prof. Dr. Bernhard Ohnesorge
Prof. Dr. Christiane Pfarrer

2. Gutachter: Dr. Maike Heppelmann

Tag der mündlichen Prüfung: 21.11.2016

Meiner Familie

Inhaltsverzeichnis

1 Einleitung ... 7
2 Ergebnisse ... 10
2.1 Publikation I: ... 10
2.2 Publikation II: ... 21
2.3 Publikation III: ... 32
3 Übergreifende Diskussion ... 41
3.1 Die Wirksamkeit von vorbereitenden Maßnahmen (Reinigung, Desinfektion) an der Injektionsstelle vor einer Injektion ... 41
3.1.1 Desinfizierende Reinigung ... 41
3.1.2 Kürzen des Haarkleides ... 42
3.2 Der Einfluss einer Kürzung der Haare auf die Entstehung von Hautstanzen und Haarpartikeln ... 42
3.2.1 Hautstanzen ... 42
3.2.2 Haarpartikel ... 43
3.3 Die histologische Zusammensetzung von Hautstanzen ... 44
3.3.1 Der Einfluss der Kanülengröße auf die Verschleppung von Hautstanzen ... 44
3.3.2 Der Einfluss der Anwendung von Kanülen mit Stilett auf die Entstehung von Hautstanzen ... 45
3.4 Die Bedeutung der semiquantitativ oder qualitativ nachgewiesenen Bakterien auf die Entstehung einer Injektionskomplikation ... 45
3.5 Übertragbarkeit der Ergebnisse auf andere Injektionsarten ... 46
3.6 Zusammenfassung ... 47
4 Zusammenfassung ... 48
5 Summary ... 50
6 Literaturverzeichnis ... 52
7 Danksagung ... 56

Abkürzungsverzeichnis

i.m.	intramuskulär
i.v.	intravenös
s.c.	subkutan

1 Einleitung

Schäden nach Injektionen sind in der Pferdemedizin seit langem bekannt. Dabei sind intramuskuläre (i.m.) Injektionen haftpflichtrechtlich Eingriffe, die der Sorgfaltspflicht unterliegen. Für Injektionen wird ein möglichst antiseptisches Vorgehen gefordert (EIKMEIER 1990 b; DEEGEN 2003; STANEK 2003), wobei eine Keimfreiheit der Injektionsstelle durch gängige Desinfektionsmethoden allerdings nicht erreicht werden kann. Vielmehr wird davon ausgegangen, dass lediglich eine Keimminderung (EIKMEIER 1990 b; STANEK 2003) bzw. eine mechanische Reinigung der Injektionsstelle erzielt wird (GERBER et al. 1980). Die Kürzung der Haare wird nur bei langem Haarkleid zur Unterstützung bei der Reinigung gefordert.

Nebenwirkungen nach intramuskulären Injektionen werden auch unter dem Begriff unerwünschte Arzneimittelwirkungen (UAW) zusammengefasst (HAPKE 1981). Diese lassen sich in schwere und leichte UAW unterteilen. Leichte UAW entstehen bei geschätzt 2,2% der Injektionen (ödematöse Schwellungen, geringgradige Hämatome, vorrübergehende leichte Bewegungsstörungen und Urtikaria). Schwere UAW werden mit geschätzt 0,045% seltener beobachtet (hochgradige Hämatome, Ödeme und Bewegungsstörungen, Phlegmonen, Abszesse, Dyspnoe, Exzitationen, Zittern, Schwitzen, Schwanken, Niederstürzen und Tod). Solche schwere UAW wurden signifikant häufiger beobachtet wenn vor der Injektion keine Reinigung durchgeführt wurde oder das Haarkleid geschoren wurde (OHNESORGE et al. 2006).

Injektionsschäden können durch physikalische und chemische Eigenschaften der Injektionslösungen, oder kontaminierte Injektionsflüssigkeiten, Kanülen oder Spritzen entstehen (AMMANN 1954). Die Bakterienflora der Haut und der Haare wurde hier als Hauptinfektionsquelle angesehen (KOLLER 1947).

In diversen Publikationen wurde nachgewiesen, dass bei der Punktion der Haut mit der Injektionskanüle regelmäßig Hautpartikel ausgestanzt werden (WILLENER 1948; LITTLE 1955; GIBSON u. NORRIS 1958; BRANDUS 1968; CAMPBELL et al. 1996; PIESOLD et al. 1998; ÖZYURT et al. 2000; PUOLAKKA et al. 2000; BARIS et al. 2004; TUNALI et al. 2006; GULDOGUS et al. 2008; ADAMS et al. 2010; BELAVY 2010; WAHL et al. 2012). Solche Hautstanzen stellen ein Transportvehikel für Mikroorganismen dar, die auf diesem Weg in das Gewebe eingetragen werden können. Die Menge des bakteriellen Eintrags soll

von der Menge der produzierten Hautstanzen abhängen (WILLENER 1948). Zwar verursacht somit fast jede Injektion einen Bakterieneintrag ins Gewebe. Es wird aber postuliert, dass diese Bakterien durch das Immunsystem unter normalen Bedingungen eliminiert werden können, so dass klinisch manifeste Infektionen nur selten entstehen (AMMANN 1954).
Bei reduzierter Abwehrlage oder einer großen Anzahl pathogener Keime kann jedoch das Verbringen von kontaminierten Hautstanzen in tiefere Gewebeschichten zu Infektionen und Abszessen führen (GIBSON u. NORRIS 1958). Diese Hypothese wird unterstützt durch Bakteriennachweise aus Kanüleninhalten nach Hautpunktionen in der Humanmedizin. So gelang es LITTLE (1955) anhämolysierende Staphylokokkus aureus auf Haustanzen nachzuweisen. BERNAU et al. (1985) konnte Mikrokokken, Propionibakterien, aerobe Sporenbildner und Clostridien nach Kniegelenkspunktionen isolieren. Auch in anderen Studien zu Gelenkpunktionen beim Pferd wurden kontaminierte Hautstanzen nachgewiesen (ADAMS et al. 2010; WAHL et al. 2012).

Eine entsprechende Vorbereitung der Injektionsstelle soll der Entstehung solcher Schäden entgegenwirken (ANDREW 1976).

Der Eintrag von Bakterien bei intramuskulären Injektionen beim Pferd in Abhängigkeit von Vorbereitungsmaßnahmen wurde bereits durch KOLLER (1947) unter Verwendung seinerzeit üblicher Mehrweg-Stahlkanülen (Durchmesser 1,5 mm) untersucht. Bei unvorbereiteter Haut betrug der mittlere prozentuale Keimgehalt auf Kanülen nach dem Durchstich 100%, nach der Schur 61%, nach Desinfektion 54% und nach Schur und Desinfektion 38%. Eine morphologisch orientierte Studie wurde von WILLENER (1948) ebenso mit Mehrweg-Stahlkanülen (Durchmesser 1,2-2,5 mm) durchgeführt. Er wies lichtmikroskopisch den Eintrag von Hautstanzen bei intramuskulären Injektionen beim Pferd nach. Den geringsten Gehalt an Hautstanzen beobachtete er bei nicht geschorenem Haarkleid, der Verwendung der kleinstmöglichen Kanüle, der Ausrichtung des Einstichwinkels in Richtung des Haarstriches, und wenn der Kanülenanschliff von der Hautoberfläche weg orientiert war. Zu dieser Fragestellung führten WAHL et al. (2012) bakterielle Untersuchungen von Hautstanzen nach Gelenkpunktion durch und konnte trotz antiseptischer Vorbereitung bei 2,61% der Punktionen in den Kanüleninhalten Bakterien nachweisen.

Ziel der vorliegenden Studie war es, den Eintrag von Partikeln und Keimen bei Injektionen mit derzeit verwendeten Injektionskanülen (Einwegkanülen; Größe: 1,2 mm ohne Stilett, 1,2 mm mit Stilett, 0,7 mm ohne Stilett) zu prüfen. Im zweiten Teil der Studie sollte der Bakterieneintrag durch Hautstanzen in Abhängigkeit von Kanülenart, Vorbereitung und Desinfektion der Injektionsstelle untersucht werden. Ergänzend sollte die histologische Beschaffenheit der Hautstanzen, deren Inzidenz und Größe in Abhängigkeit von Kanülenart und Vorbereitungsmethode der Injektionsstelle untersucht werden.

2 Ergebnisse

2.1 Publikation I:

Complications following intramuscular injections in equids

Published in: Journal of Equine Veterinary Science

Volume 35, Issue 6, June 2015, Pages 465–474

Received: November 12, 2014; Received in revised form: January 26, 2015; Accepted: February 1, 2015

T. Puschmann, B. Ohnesorge

Clinic for Horses, University of Veterinary Medicine Hannover, Foundation

Bünteweg 9, 30559 Hannover, Germany

Abstract:

Literature on complications following intramuscular injections in equids is fairly sparse and consists of a few studies, case series, personal opinions, and surveys. The necessity of injection site preparation protocols and appropriate localization for intramuscular injections are often discussed, especially when compensation claims arise. The variety of adverse reactions and their symptoms is well-known to all practitioners, with some being less frequently observed than others. Especially the contribution of tissue irritating capacities of intramuscularly administered drugs to abscess formation or even clostridial myonecrosis is discussed. With this review, we aim to summarize the current knowledge, standards and opinions on injection site preparation protocols, injection techniques, adequacy of injection sites, complications following intramuscular injections, and possible causes of intramuscular injection complications.

Journal of Equine Veterinary Science 35 (2015) 465–474

Contents lists available at ScienceDirect

Journal of Equine Veterinary Science

journal homepage: www.j-evs.com

Review Article

Complications After Intramuscular Injections in Equids

Tobias Puschmann*, Bernhard Ohnesorge

Clinic for Horses, University of Veterinary Medicine Hannover, Foundation, Hannover, Germany

ARTICLE INFO

Article history:
Received 12 November 2014
Received in revised form 26 January 2015
Accepted 1 February 2015
Available online 11 February 2015

Keywords:
Intramuscular
Injection
Complication
ADR
Infection
Horse

ABSTRACT

Literature on complications after intramuscular (IM) injections in equids is fairly sparse and consists of a few studies, case series, personal opinions, and surveys. The necessity of injection-site preparation protocols and appropriate localization for IM injections are often discussed, especially when compensation claims arise. The variety of adverse reactions and their symptoms is well known to all practitioners, with some being less frequently observed than others. Especially, the contribution of tissue-irritating capacities of IM administered drugs to abscess formation or even clostridial myonecrosis is discussed. With this review, we aim to summarize the current knowledge, standards and opinions on injection-site preparation protocols, injection techniques, adequacy of injection sites, complications after IM injections, and possible causes of IM injection complications.

1. Introduction

Intramuscular (IM) injections are a common technique used for administering drugs in equine medicine. It is concluded that this application technique is easily performable and effective for drug administration [1]. Complications after IM injections are estimated to show the highest prevalence in equids [2–4], and sometimes occur even if injections are performed accurately [5]. When performing injections, veterinarians have a contractual relationship toward the animal's owner and are responsible for correct execution, but subsequently, do not guarantee the outcome [6]. Although fatal outcomes of adverse drug reactions (ADRs) after IM injections are observed infrequently with 0.0015% of cases, and severe ADRs occur with 0.045%, low ADRs are reported with an incidence of 2.2% [7]. Because complications after injections can be lethal, they sometimes engender compensation claims by the owner toward the treating veterinarian [2]. Intramuscular injections are considered as subject to the duty of care and fall under legal liability. Preconditions for performing IM injections are an indication, the agreement of the animal owner, and the completion exercised with due care [8]. An indication for IM injections is given when it is the most promising and risk-free method for drug administration. Thus, an oral or rectal application has to be favored if the therapeutic effect is the same [9]. If injections are not performed with due care (lege artis), the executing veterinarian can be held responsible for possible complications [6]. Judicially, the veterinarian in charge has to choose a treatment which suits the patient's condition, with adequate medication in reasonable doses, and the site and route of administration has to be adequate for the pharmaceutical chosen. Informing the owner about adverse reactions is necessary when certain risk-carrying pharmaceuticals, such as benzathine-penicillin, are administered, because lethal systemic adverse reactions must be anticipated [5]. Discrepancies can be observed between the actual practical work of veterinarians and textbook procedures concerning IM injections, leading to the questioning of whether the negligence of preparation methods was etiologic for the occurrence of adverse reactions [10].

* Corresponding author at: Tobias Puschmann, Clinic for Horses, University of Veterinary Medicine Hannover, Foundation, Bünteweg 9, 30559 Hannover, Germany.
E-mail address: tobias.puschmann@tiho-hannover.de (T. Puschmann).

0737-0806/$ – see front matter
http://dx.doi.org/10.1016/j.jevs.2015.02.001

2. IM Injections

2.1. Preparation and Disinfection of the Injection Site

Before performing IM injections, the injection site must be carefully examined and palpated to identify any contamination or/and tissue alteration, such as dermatitis and cellulitis, to guarantee that injections are not carried out in areas with pathologic tissue abnormalities [9]. Every single injection should be performed as antiseptically as possible [9], and disinfecting or cleaning the injection site must be carried out before each injection [2–5,11,12]. Sterile conditions will not be achieved by a single disinfection [3,9]. Nevertheless, a mechanical cleaning of the injection site will be accomplished [6] together with a reduction of the microbial population [9]. If disinfection is not performed, or performed insufficiently, before injections in horses, it can be considered a negligent or grossly negligent act [9]. Contrarily, it is considered sufficient in human medicine to perform injections without swabbing if the skin at the injection site is clean. The injection site must be washed before injections are performed only if the localization is visibly soiled [13].

However, more than half of equine practitioners disinfect the injection site, only 0.46% use clipping as a preparation method [14]. The clipping of the hair at the injection site is only considered necessary if the hair is long [3,5,15]. Shaving the hair is not an option because it produces more reactions of the skin, which leads to a greater risk of infection [16]. Shaving leads to more intense defense reactions when subcutaneous injections are performed and causes the most intense inflammatory reactions at the injection site [10].

2.2. Materials and Fluids Used for IM Injections

All instruments used for IM injections must be sterile [3,5,6,9,12,13]. When sterile materials are used for IM injections instead of resterilized materials, the incidence of an ADR is lowered significantly [7]. Inspecting the packaging of materials for integrity before usage is advised [13].

The needle size should be adjusted to the size of the horse [1] since pain reactions, the incidence of bleeding after injections, and tissue reactions after injections increase with larger needle sizes [10]. The size must also be adjusted concerning the injection volume and the viscosity of the pharmaceutical administered [6]. Comparing the intensity of postinjection inflammatory reactions to different needle sizes, the intensity of the inflammatory reactions correlates positively with the size of the needle [10]. Rose and Hodgson [17] use 3.75-cm long, 18- or 19-ga needles for IM injections, whereas Rossdale and Ricketts [18] prefer 19- or 21-ga needles, with a length of 3 cm in adult horses and 2 cm in foals.

Drugs administered IM should have good tissue compatibility and must be accredited for this application route. Drugs must also be administered in a harmless composition and amount [5] and must be sterile and without precipitates [19]. Combinations of pharmaceuticals in one syringe can derogate their therapeutic effect, as well as their local and systemic compatibility [9]. If disinfecting the top of drug vessels before extracting the solution is not performed, low ADRs will occur significantly more often [7]. In human medicine, the use of new sterile needles for every puncture of multidose vials, which should be removed immediately afterward, is advised [13]. Very rarely, the drug itself can carry pathogenic germs [2]. The multiple puncturing of pharmaceutical closures of drug vessels may lead to a contamination of the content with bacteria, with the risk being especially high in vaccination sera because they represent ideal culture media for bacteria [2,3]. The volume deposited at one injection site should not exceed 20 mL [9] to 30 mL [5]. However, a survey indicates that 68% of veterinarians administer 20–25 mL per injection site, whereas 9% even administer more than 30 mL [7].

2.3. IM Injection Technique

Antiseptic hand washing is not demanded for performing IM Injections, but hands should be generally washed before applications [3]. The injection site should be tapped with the hand a few times before needle insertion [17]. If movement of the animal is expected [12], or in general, inserting the needle separately and attaching the syringe afterward are advised [18]. Berge and Westhues [12] advise only inserting the needle for two third of its length to avoid its breaking off. Contrarily, Rose and Hodgson [17] state that needles must be inserted up to the needle hub, for the position of the needle may change during the injection, possibly leading to an inadvertent intravascular injection. After inserting the needle into the muscle, it is important to check, by drawing back the plunger, whether it has punctured a vessel [1,5,9,17,20]. It is considered insufficient to mainly watch the needle hub for emerging blood drops [1]. Low ADRs are also observed significantly less frequently if aspiration testing for inadvertent vascular puncture is performed [7]. If a vessel is punctured, another injection site must be used [1,9]. In this case, it is also necessary to make use of new needles as blood clots can plug the needle lumen, thus leading to false-negative aspiration testing. If larger volumes have to be administered, Berge and Westhues [12] advise spreading the solution by repositioning the needle within the muscle tissue. Secondary bleeding from the injection canal is usually harmless and stagnates after short manual compression [9]. Koller concludes that massaging the injection site after the injection, to enlarge the region where resorption is achieved and also to increase the resorption rate by applying pressure to the injection depot, is recommended. The exposure time of the surrounding tissue toward the partly damaging effects of pharmaceuticals could be reduced in this way [4]. If IM injections have to be performed repeatedly in the same patient, it is suggested that different injection sites should be used [9,17,21]. Alternating left and right sides of the horse also help to keep inflammatory reactions at the injection sites as minimal as possible [17]. Even pharmaceuticals which only have mild tissue-damaging effects can cause severe tissue alterations if administered regularly at the same injection site [22]. Deegen [5] advises the veterinarian to stay with the patient for at least 5 minutes after the injection to be able to intervene in cases of emerging adverse reactions.

2.4. Localizations Used for IM Injections

To guarantee an injection performed lege artis, an adequate injection site must be chosen [5]. Such localizations are supposed to contain large muscle masses, be poor in tendons, and close to the surface, to guarantee a fast resorption. They should not be next to larger vessels or nerves. In case of complications after the injection, the injection sites should be easily and safely accessible for therapy and should cause the patient as little discomfort as possible [23]. Injection sites should also be chosen regarding the volume and the kind of pharmaceutical to be injected [1].

2.5. The Pectoral Musculature

The pectoral musculature is considered an appropriate localization for IM injections [5,17,22,24]. A survey indicates that 33% of IM injections are given into the pectoral musculature [7]. It is indicated that pain originating from injection complications at this site may cause pain-associated problems in stallions, when covering mares [24]. This injection site is considered suitable for the injection of small volumes, although abscessation is observed more frequently than in other injection sites [1]. Nevertheless, in cases of abscessation, surgical therapy at this location is considered easy to perform. Injections performed too close to the sternum are hazardous because they may cause damage to the sternum itself, leading to fistula formations [24]. Injections into the pectorals, especially in young horses, are not risk-free procedures, and tissue irritations may cause gait abnormalities [25]. Gerber et al [6] observed more tissue reactions when IM injections were given into the pectorals than in the neck, concluding that IM injections should not be performed at this site. Garbade observed subcutaneous swellings after IM injections into the pectoral musculature relatively frequently and discusses whether it is because of drug reflux via the injection canal. The healing process of these lesions, nevertheless, was first initiated in the pectorals, probably as a consequence of the good muscle perfusion [22].

The incidence of severe ADRs after IM injections increases with high significance, if injections are given into the pectorals compared with the hamstring area, the gluteal region, and the neck. Vaccines administered in the pectorals cause ADRs even more frequently with high significance compared with the hamstring region [7].

2.6. The Hamstring Musculature

The caudal hamstring area is also considered an adequate injection site for IM injections [5,17,22]. However, only 14% of IM injections are administered at this location [7]. Injections performed too far laterally or distally may end up in the musculus biceps fermoris, compromising bigger veins or sciatic nerve branches [26]. The safety risk to which veterinarians are exposed is disadvantageous when performing injections at this site [1,25]. Subcutaneous swellings, as frequently observed when injecting into the pectoral musculature, were not observed when injections were administered in the hamstring area [22]. Nevertheless, gait abnormalities due to tissue irritations were observed after IM injections at that location [25]. Furthermore, the possibility of lameness or iatrogenic bleeding is mentioned [26]. In cases of abscessation, drainage can be achieved fairly well. The anatomic proximity to the sciatic nerve provides the risk of sciatic paralysis in foals, especially when poorly muscled [1].

2.7. The Neck Musculature

The adequacy of the neck muscles as an IM injection site is discussed ambiguously in literature. Although Deegen [5] considers the neck muscles to only partly be adequate for IM injections, Garbade [22] and Wissdorf et al [27] consider it suitable as an injection site. The neck muscles are, with 47%, the most frequently used IM injection site in horses [7]. Boyd studied IM injection sites in the neck using contrast medium administrations. Intramuscular deposition of contrast medium was most successfully achieved when injections were performed dorsal of the upper line of the musculus brachiocephalicus and ventral of the ligamentum nuchae, at the level of the fifth vertebra. Nevertheless, even at that location, contrast medium was infrequently deposited into deeper fascias of the extensory musculature of the spine. With all other sites examined, contrast medium was deposited close to the ligamentum nuchae, the vertebrae or fascias, which in conclusion, may cause severe complications, especially if injected materials tracked along these fascial sheaths [28].

One author concludes that only smaller volumes should be administered into the neck because the muscle mass is not big enough and neighboring structures can be compromised [1]. Furthermore, it is stated that the anatomic conditions recommended for IM injection sites are not fulfilled in the neck. Therefore, it was not the most risk-free method which had to be chosen according to due diligence, if there was no specific reason for choosing this injection site [23]. Because the muscles in the neck area are fairly thin and separated by fascias [6,23], a restriction of expanding tissue reactions can lead to subsiding reactions or cause pressure-implied damage to neighboring structures, such as the esophagus, Vena jugularis, Nervus vagus, and Nervus sympathicus [23]. Considering complications after IM injections in the neck, some authors state that they were only mild and occur very infrequently [6,25]. Contrarily, Eikmeier [23] states that movement abnormalities caused by injection complications in the neck can present themselves substantially because of the inability of the tissue reactions to expand at this location. Kollowa [29] also observed substantial reactions after vaccinations in the neck musculature and advises injecting irritating drugs into the pectoral musculature in horses.

If IM injections are given in the neck, the incidence of ADRs is significantly increased compared with IM injections in the hamstring region. Adverse drug reactions occur more frequently with high significance especially when vaccines are administered in the neck compared with the hamstring area [7]. Surgical intervention in this area can be fairly difficult if abscessation occurs [1,23,27], or can even be impractical [23]. The neck musculature in foals is not developed enough to allow IM injections, which leads to a higher risk of compromising neighboring structures [1].

2.8. Other Localizations Used for IM Injections

Intramuscular injections in the gluteal region are considered critical by Wissdorf et al [30] because abscessation may subside deep into the muscle tissue and cause gait abnormalities. Deegen [5] even considers it illegitimate because infections can cause irreversible damage at this localization, whereas Williams [1] and Rose and Hodgson [17] state that it was optional as an injection site. Williams considers it especially for larger injection volumes because the muscle bellies are relatively large. However, he also mentions that abscessation drainage is difficult in this area [1]. Intramuscular injections in the gluteal region are administered in 5% of cases, whereas the least frequently used injection site is the musculus triceps brachii with 1% [7]. Deegen [5] and Wissdorf et al [31] consider the musculus triceps brachii to be an adequate injection site for IM injections, and Rossdale and Ricketts [18] consider it usable in very young foals. The musculus biceps femoris is only an optional IM injection site in foals, as it is a relatively large muscle at this age [1].

3. Possible Complications After IM Injections

3.1. Complications Originating From the Injection Procedure and Sterile Tissue Alterations

A possible complication originating from the injection procedure is the breakage of needles. The executing veterinarian will only be held responsible if the patient was not restrained correctly or if the material was deficient [9]. Hematomas can infrequently develop from puncturing vessels, but are considered a harmless complication in human medicine. Nevertheless, it is considered contraindicative to perform IM injections in patients experiencing coagulopathies or patients being treated with anticoagulants [32]. Small hematomas occur in equids in 0.33% of IM injections, whereas larger hematomas only occur in 0.005% [7].

Aseptic tissue necrosis may develop at the injection site, supposedly when incompatibility reactions occur toward the injected solution [32,33], independently of injection techniques [33]. Scar tissue and fibrosis may lead to functional and cosmetic ailments [32]. Aseptic tissue necrosis was frequently observed in puppies after IM injections [33]. In human medicine, severe aseptic tissue necrosis after IM injections is also described as “embolia cutis medicamentosa” or “Nicolau syndrome” [34,35]. This syndrome in men is supposed to originate from intra-arterial injections or injections next to arteries after arterial puncture [36–38].

Nerve injury may occur after IM injections and is well described in humans, mainly concerning sciatic nerve injury [39,40]. Etiologies discussed are scar tissue formation in and around the nerve [41], injections of neurotoxic solutions intraneurally or paraneurally [40], and direct injury of the nerve through the needle or the pharmaceutical injected [42]. Nerve compression through injection volumes, intraneural and paraneural hematomas, edema formation, tissue ischemia [43], and allergic nerve reactions after injections [44] are also discussed.

Solely edematous tissue reactions also occur after IM injections and can be characterized as a symptom of ADRs, differentiating between severe edema with an occurrence of 0.005% and slight edema with 1.56% of IM injections [7]. Hindrances in movement occur due to tissue reactions after IM injections and can present themselves severely [23]. As a symptom of ADRs, severe handicap in movement occurs in 0.02% of IM injections, whereas slight short-term movement abnormalities are observed more frequently in 0.19% [7].

3.2. Adverse Drug Reactions

It is stated that all pharmaceuticals have potentially harmful effects. It is declared that pharmacologically speaking, side effects cannot be differentiated from the main effects. Pharmaceuticals rather include certain properties that can be used therapeutically, whereas others are unwanted [45]. A separation of the requested and unrequested effects is impossible for they originate from the same effect mechanisms [46]. Thus, it is recommended to redefine these side effects as ADRs [45].

Severe ADRs occur in 0.045% of IM injections annually, whereas low ADRs occur in 2.2% of IM injections annually. Severe ADRs occur significantly less with practitioners who frequently perform IM injections [7]. Adverse drug reactions include intolerances against pharmaceuticals [46], toxic reactions, idiosyncrasia, anaphylactoid reactions and anaphylactic reactions [46,47], intolerance reactions of the skin, and sudden death [47]. Urticaria, considered a low ADR symptom, occurs in 0.09% of IM injections and appears usually 1–30 minutes after the injection [48]. Regarding the other systemic symptoms rated as severe ADRs, shaking and sweating occur most frequently in 0.005% of IM injections annually, followed by swaying (0.004%), dyspnea (0.002%), and excitations and collapse (0.001%) [7].

Considering the pharmaceuticals causing severe ADRs, most of the local reactions were registered after administration of vaccines (57%), followed by antibiotics (22%) and vitamins (9%). Most of the systemic reactions occurred after administration of antibiotics (51%), followed by vaccines (20%) and vitamins (19%). Overall, all pharmaceuticals caused more local than systemic reactions [48]. Eighty-one percentage of ADRs after IM antibiotic administration occurred after injections of drugs containing benzylpenicillin, including nearly all lethal ADR cases. Lethal ADRs after IM injections occur in 0.0015% of injections annually and appear seconds to 30 minutes after the injection [7]. Almost all lethal ADRs reported followed IM injections of antibiotics and vitamins [48]. Some individual cases are reported where lethal ADRs followed injections of vaccines, sedation, and homeopathic drugs [7].

Considering ADRs, a complete prevention is not considered achievable because every administration of a pharmaceutical carries a certain degree of risk. Calculating the value of the pharmaceutical effect compared with the risk of a potential ADR is the veterinarian's duty, as well as keeping the risk of complications as low as possible. In addition, it is demanded to make use of pharmaceuticals in exact doses, to avoid ADRs through toxic reactions, and also an avoidance of the combination of various pharmaceuticals of which the interaction pattern is unknown. Because anaphylactoid and idiosyncratic reactions are unpredictable, they cannot be considered medical malpractice [45]. Drug applications through injections should be carried out only if necessary [46,47,49], even more so when pharmaceuticals are supposed to have sensitizing effects [46]. If ADRs occur frequently with one product, it is advisable to reconsider its therapeutic use [47].

3.3. *Infections Originating From the Injection Site*

Ammann [2] states that bacteria from the skin are introduced into the deeper tissue layers with almost every injection. Furthermore, pathogenic bacteria can be carried along, via skin cores or contaminations, and are introduced into the deeper muscle tissues along with the needle insertion [10]. However, this is considered not to necessarily lead to clinically manifest infections of the injection site because physiologically infectious agents can be eliminated. Furthermore, another etiologic possibility for injection-site infections is mentioned, stating that bacteria may gain access to the injection site on the endogenous path by using the blood stream [2]. The main infective complications after injections are mentioned to be malignant edema, followed by inflammatory processes, leading to cellulitis and abscess formation [19]. Cellulitis and abscesses both occur in 0.016% of IM injections annually [7]. The etiology of injection abscesses in human medicine is considered an either primary or secondary injection-site infection. Infective agents may gain access to the injection site by an exogenic or endogenic pathway, with the latter being subordinate. Injection abscesses are more frequently observed in immunocompromised humans, with very infrequent complex or lethal progression due to hematogenous metastasis [32].

Zeller [3] summarizes that the chemical characteristics of pharmaceuticals, such as deviations of pH values from the pH values of the surrounding tissue, may cause necrotic tissue alterations at the injection site. The retention period of pharmaceuticals at the injection site, as described by Garbade seems to correlate with the incidence of muscle necrosis and degenerative tissue alterations, as well as the incidence of bleeding in the musculature. These necrotic areas are discussed as an initiative for cellulitis or abscess formations [22]. Injection fluids may be contaminated with bacteria and be a cause for injection-site infections [2]. In two case series concerning clostridial myonecrosis, no connection whatsoever could be drawn to contaminated pharmaceuticals [50,51]. Nevertheless, contamination of pharmaceuticals causing injection abscesses is described in human medicine [52,53].

Rapidly progressing cellulitis and myositis occurring after IM injections are frequently caused by clostridia, with the exact pathogenesis of these infections remaining unknown [54,55]. Clostridial myonecrosis in horses developing from causes other than trauma are not described in literature. In the study from Peek et al [50], 34 of 37 horses presented with clostridial myonecrosis after soft-tissue injections, and only three horses had a history of initial injury. Other publications show similar results [51,54,56]. Of practitioners, 6.4% state having encountered at least one case of clostridial myonecrosis after IM applications [14]. Monoinfections with *Clostridium perfringens* occur most frequently, showing the highest survival rates [50]. Similar results are reported in other cases with the majority or all patients surviving [54,56,57].

In addition, monoinfections with *Clostridium septicum* are described, reporting lethal outcomes in most studies [51,54,56,58]. The only case series reporting surviving cases is published by Peek et al [50] with a 50% survival rate.

Further reports include lethal monoinfection with *Clostridium chauvoei* [54] and *Clostridium botulinum* type B [59]. Furthermore, clostridial myonecrosis is reported to also occur as mixed infections [50,51].

Clostridia are expected to be located on the skin, the needle, on the drug vessel, or in the drug itself [14]. It is described that dormant clostridia spores in cattle can be located in the musculature, which can be reactivated by traumatic influences and then start to multiply [60]. Also, in horses, some clostridia species have been isolated out of healthy muscle tissue. A reactivation through traumatic influences such as IM injections is discussed [61]. Brown et al [14] suggest that absorption of bacteria and spores from the intestine may be promoted by colic or parasitic stages after IM ivermectin application because of inflammatory processes. An experimental study on guinea pigs indicates that endotoxins can favor the translocation of intestinal bacteria into other organs [62]. It is discussed that, in this context, tissue necrosis after IM injections creates ideal conditions for the accumulation of anaerobic bacteria [14,61].

The general survival rate of clostridial myonecrosis is described as 73%, whereas monoinfections with Clostridium *perfringens* have better outcomes with 81% [50]. To avoid the risk of clostridial infections after IM injections, Brown et al [14] suggest avoiding IM injections of nonantibiotic solutions if other application routes, for example, PO or IV application, are accredited. Because 24 of 34 horses with clostridial myonecrosis were injected in the neck in the study of Peek et al [50], the conclusion is to administer tissue-irritant drugs IV if accredited, or deep into the well-perfused hamstring area if only accredited for IM use, as the incidence was lower at this injection site, and well-perfused muscle tissue is essential for minimizing anaerobic infections. Brown et al [14] conclude by comparing the application methods and preparation protocols with the incidence of clostridial myonecrosis that there is no significant correlation to be found.

An ongoing discussion exists on IM administration of flunixin as a cause of clostridial myonecrosis, especially in the United States, on nonscientific web pages. Banamine (flunixin meglumine) is accredited for IM use in horses [63]. Nevertheless, very little scientific literature can be found considering this issue. There are several reports of clostridial myonecrosis after IM injections of flunixin [50,54,56,64,65], as well as one case of botulism [59]. In a study performed on six horses, comparing flunixin disposition after IM administration, no injection-site reactions were observed [66]. Nevertheless, repeated IM flunixin applications are not advised because complications such as muscle soreness, abscess formation, and/or myositis may occur [67].

4. Possible Causes of Complications After IM Injections

4.1. Injection-Site Preparation Protocols and Injection Material

The most successful preventive method against injection complications would be an aseptic procedure for IM injections [51]. However, even the latest skin disinfection methods are incapable of completely sterilizing the skin from bacteria. Therefore, it is concluded that prevention of infections derived from the injection site should be pursued by proper disinfection [68]. The preparation method of the injection site is stated to only be one of many factors influencing the complexity of complications after IM injections. Considering the inflammatory symptoms of the injection site itself after IM injections, they appear most distinct when injection sites are shaved and disinfected, followed by nonclipped and nondisinfected injection sites and clipped and disinfected injection sites. In addition, bigger needle sizes produce more severe tissue lesions at the injection site than smaller ones [10].

Koller studied the influx of bacteria via the needle when puncturing skin preparations, comparing different preparation protocols. If the mean percentage of bacterial content of the needle when puncturing nonclipped and nondisinfected skin was set at 100%, it was reduced after solely clipping to 61%, after solely disinfecting to 54%, and the greatest reduction was achieved after clipping and disinfecting to 38%. Consecutively, he concludes that, from a microbiological point of view, clipping is a reasonable method for bacterial reduction [4]. Contrarily, Hague et al [69] published results of skin preparations of the midcarpal joint, indicating that clipping before antiseptic preparation increases colony-forming units (CFUs) significantly compared with nonclipping. Nevertheless, Willener [10] was able to show that clipping increases the frequency of skin coring. Considering this, Koller [4] concludes that the benefit of bacterial reduction through clipping may be relativized by the more frequent incidence of skin cores.

Additionally, Zubrod et al [70] published results indicating that there is no significant difference in CFUs after antiseptic preparation of contaminated and clean skin, concluding that antiseptic site preparation can only minimize the risk of potential infections. If one had to guarantee sterile conditions for IM injections, the injection site would have to, nevertheless, be clipped and antiseptically prepared, which is considered to be impracticable or unnecessary because of the infrequent occurrence of adverse reactions [1]. Considering the relevance of clipping for antiseptic preparation, no significant difference in CFUs comparing clipped versus nonclipped preparation sites postscrub was verified [69].

The bacterial skin flora is considered the main source of infections after IM injections, which is why bacterial reduction methods must be considered as the highest priority [4]. It is concluded that, although it is known that bacteria and skin cores are still translocated into the deeper tissue layers with disinfection before injections, preparation protocols should not be ignored [2]. Although it seems unlikely that clostridia, which are highly resistant, are eliminated by disinfection with alcohol, the cleansing effect of this method should be considered [56]. Evaluation of the occurrence of ADRs after IM injections shows that the incidence is significantly greater when injection sites are clipped or are not cleaned [7].

4.2. Mechanical and Chemical Effects of Drugs at the Injection Site

Injected solutions may damage the surrounding tissue, and thus, reduce the local physiological immune defense mechanisms through their chemical and mechanical effects at the injection site [2]. The tissue is suggested to tear around the injection depot, and vessels may be condensed because of rising injection volumes or too fast an injection speed [4]. Contrarily, it is shown that slight bleeding and fiber disruption occurs independently of the injection volume, and the muscle damage does not correlate with the injection speed [71]. Furthermore, the possibility of pharmaceuticals causing vascular constriction is discussed, which may promote infections with anaerobic bacteria. Also, hypertonic solutions are discussed to cause edema formations at the injection site through dehydration of the surrounding tissue, leading to autolytic cell degradation [4]. Furthermore, oily solutions are supposed to cause edema formation at the injection site as well [2]. These alterations, intensified by rising injection volumes, are thought to reduce the ability of local immune defense mechanisms to access these areas [4].

The local tissue compatibility of medicinal products after IM injections correlates with various factors, such as pH value, concentration, effective osmolarity, volume, irritation potential, and the injection rate per time unit [72]. It is suggested that greater pH value deviations from the surrounding tissue may cause tissue necrosis [4]. A study indicates that pH values of pharmaceuticals do not correlate with the development of muscle lesions if only pH values of 3–10 are considered. Considering the effect of the osmotic ratio on muscle lesions, pharmaceuticals with an osmotic ratio (1 = osmotic pressure of saline solution) of more than five caused tissue lesions. Sixty-nine percentage of pharmaceuticals caused cytotoxic cell alterations to various degrees, including all antipyretic and analgesic drugs, and 34 of the 37 antibiotics tested. Furthermore, many pharmaceuticals tested feature hemolytic potential, which correlates with severe muscle lesions at the injection site [49].

When macroscopically and microscopically examining injection sites 7 days after the injection of pharmaceuticals in rabbits, 51% of the pharmaceuticals tested showed severe muscle lesions, whereas no lesions were observed in 27%. All the 16 antipyretic and analgesic drugs tested and two thirds of the 31 antibiotics tested caused severe muscle lesions. Rasmussen and Svendsen examined injection sites after IM injections of chemotherapeutics in pigs. Macroscopically, injection sites showed necrotic areas with hemorrhagic peripheral rim formations 6 days after the application. Those symptoms were absent when only saline solution or sterile water was administered [73]. The healing process depends on the intensity of the muscle damage. Marked necrosis turns into encapsulated scar formation, whereas slight myodegeneration can be regenerated completely [74].

Studying the local tissue compatibility of pharmaceuticals after IM injections via the elevation of the creatinine phosphokinase (CK) in the blood is a procedure which does not require euthanasia to examine injection sites [72]. It is a noninvasive and quantitative method to study muscle lesions after IM injections [75]. The quantity of CK elevation depends on the extensiveness of the lesions and the amount of damaged muscle tissue. The damage is caused by the pharmaceutical agent, the carrier substance, and the concentration and volume of the solution administered [76]. After IM injections of a 8.8 mg/kg of a 20% phenylbutazone solution in horses, a significant CK elevation was noticed, whereas IV injections of the same amount did not provoke a CK elevation [77]. Furthermore, a direct proportionality between damaged tissue and the injection volume was demonstrated, when injection volumes of 0.25 mL and 1 mL in rabbits and 0.5 mL and 3 mL in pigs were considered [78]. Verifying the tissue compatibility of drugs before use to minimize complications after IM injections is advised [22]. Also, the use of solutions with the least tissue-irritating capacities is advised, if one can choose between several pharmaceuticals [20].

4.3. Skin Cores Produced by Injection Needles

Skin and hair fragments are translocated into deeper tissue layers with every injection [10]. Numerous studies describe the incidence of skin cores depending on the size and type of needle, respectively [79–89]. Skin cores are discussed as being causative for infections after injections [79,81] and are considered unavoidable [81]. It is considered possible that bacteria are translocated into the deeper tissue layers adherent to skin cores [11,81]. It is even stated that skin cores should be considered the primary carriers of pathogenic bacteria because pharmaceuticals are produced sterilely, concluding that, as long as injection sites are not sterile, the rate of complications will depend on the translocation of foreign debris, skin, and hair fragments [10]. In human medicine, micrococci, propionibacteria, *Bacillus*, and clostridia were isolated from skin cores obtained by puncturing disinfected knee joints [11,90]. Wahl et al [91] isolated bacteria in 2.6% of skin cores obtained through intra-articular puncture of horse joints after antiseptic site preparation. Koller [4] also detected bacteria when culturing needles after puncturing skin preparations of horses.

It is assumed that even if skin cores were carriers of pathogenic bacteria, it would not necessarily cause complications in well-perfused muscle tissue. Nevertheless, it is considered possible that if pharmaceuticals with tissue-irritating effects are administered, and local or systemic immune defense mechanisms are, thereby, reduced at the injection site, skin cores carrying pathogenic bacteria may initiate infective complications after IM injections [81]. It is also considered possible to translocate bacteria adhering to the needle's surface because the skin will not be held sterile by the common disinfection methods [11].

The incidence of skin cores in horses is described as being greater when injection sites are clipped compared with nonclipped injection sites [10,92]. The highest incidence of skin cores, nevertheless, is detected when injection sites are shaved, whereas the amount of hair fragments found in the injection canal is highest when injection sites are clipped [10]. Furthermore, Willener discusses that skin cores must be estimated to carry a greater bacteriologic risk than hair fragments because they consist of hair follicles, sebaceous glands, and sweat glands. Bacteria are suspected of adhering more easily to irregularly shaped skin cores than to hair fragments. He concludes, moreover, that clipping and shaving are inadequate preparation procedures [10].

4.4. Other Diseases Promoting Injection Complications

Infective diseases, deficiency diseases, and weakness in general may reduce the cellular and humoral immune defense mechanisms [4]. It is advised to consider this before performing injections [19]. Concerning the incidence of injection complications, a study describes 48 of 68 animals with complications after IM injections, which were suffering from mild to severe diseases. It is suggested that local and systemic malperfusion may promote the bacterial infection of injection sites. Malperfusion may also be promoted through lung diseases causing decreased arterial oxygenation. Moreover, it is mentioned that primary and secondary anemia are causative following the same mechanisms. Liver and kidney diseases as well as gynecologic diseases are also discussed as being predisposed to injection complications [4]. Colic is the most frequently reported initial symptom of patients developing clostridial myonecrosis [50,54]. Koller reports that 30% of horses developing clostridial myositis also showed colic symptoms. However, the connection between colic and clostridial myonecrosis remains unclear [55]. Peek and Semrad [55] suggest that colic patients rolling in the dirt possibly contaminate themselves with clostridia, which then can be advanced into the muscle tissue through IM injections with analgesics. Perdrizet et al [64], on the other hand, discuss colic as being a prodromal stage of clostridial myonecrosis.

5. Conclusions

Complications after IM injections occur infrequently, which makes clinical studies difficult. Recent specific scientific literature on IM injections in equids is very sparse, and some early literature does not reflect today's scientific or practical standards. Therefore, conclusions can be drawn on several topics, whereas others certainly need further scientific investigation. Successfully summarizing new scientific results therefore cannot be achieved with this literature review. Instead, we summarize today's scientific standard and aim to highlight areas of interest where further scientific research is desirable.

For equine medical standards, performing a cleaning or disinfection of the injection site before performing injections is considered obligatory, whereas clipping is only advised if the hair is long, despite the lack of scientific literature on the difference of disinfection of clipped and nonclipped injection sites of horses. There is an early study describing that the bacterial detection of injection needle's contents after disinfection of the injection site is reduced, and even more so, if injection sites were clipped and disinfected [4]. Considering antiseptic site preparation, the antiseptic effect does not differ between unclipped and clipped skin specimens [69]. Only very infrequently drugs may contain infective agents today. Nevertheless, it is advisable to handle drug vials correctly and disinfect the top of drug vessels before extracting the solution, because it reduces the risk of low ADRs significantly [7]. Injections should be performed with small needle sizes because it reduces the intensity of inflammatory reactions at the injection site [10]. To the author's knowledge, there are no scientific studies on the maximal volume to be administered per injection site, as well as studies comparing injection volumes to different injection sites. If repetitive IM injections need to be performed, using different injection sites is advised because otherwise severe tissue alterations have to be anticipated as shown by Garbade et al [22]. The hamstring area, the neck, and the pectorals are the predominant injection sites used for IM injections [7]. In the hamstring area, complications are low and the safety risk is considered high, whereas the complication rate in the pectorals is higher. Nevertheless, alterations heal faster in the pectorals than the neck and the hamstring area [22]. The suitability of the neck as an IM injection site is ambiguously discussed. Tissue alterations are observed less in the neck than they are in the pectorals, but the risk of causing extensive damage to neighboring structures in the neck must be estimated to be the highest. It must be considered the veterinarian's responsibility to know about the benefits and risks of the different injection sites. The veterinarian should choose the most adequate injection site, considering his own safety according to the horse's character, the risk of complications causing further damage, the contamination status of the hair coat, the drug volume, and the tissue-irritating capacity of the drug to be administered.

The incidence of nerve injury and aseptic tissue necrosis is, to the best of the authors' knowledge, not described for equids in scientific studies, whereas hematomas after injections occur infrequently. The predominant complications after IM injections are edema, cellulitis, and abscess formation, often causing hindrance in movement [7]. Septicemia after injection abscess formation with severe or lethal progression is only described in humans. Clostridial myonecrosis develops infrequently after IM injections, presenting the most severe infective injection complication in horses with often lethal outcomes. Infections may be initiated through reactivation of dormant spores, which were detected in equine muscle tissue, but so far their relevance is not confirmed scientifically [61]. Another possibility is the transport of bacteria adhering to skin cores or debris, carried into the muscle tissue along with injection needles, because even after antiseptic site preparation, bacteria were detected in needle contents after skin punctures, including clostridia in two studies in men [11,90,91]. Nevertheless, the tissue-irritating capacities of drugs must be considered to contribute to infective complications as well, and their accreditation for IM use might need reconsideration. A large-scale field evaluation of complication rates due to tissue-irritating capacities of flunixin, for example, after IM injections are desirable, as it is discussed on Internet forums and also Web pages of veterinary services. Measuring the CK elevation after injections is a valuable method for studying tissue-irritating capacities of drugs [77].

Complications after IM injections occur very infrequently. Nevertheless, some complications turn out severe or even cause fatalities. There are only very few scientific studies considering this topic. Larger multicenter studies are desirable to furthermore reduce complications and adverse reactions or to evaluate the suitability of drugs for IM injections and the influence of tissue-irritating capacities on infective complications. Therefore, reporting and evaluating ADRs is obligatory to every veterinarian. Although anaphylactic, anaphylactoid, and idiosyncratic reactions have to be considered unforeseeable, infective complications can be reduced. Therefore, further investigation of a possible exogenic or endogenic path of infection is needed, especially for minimizing the incidence of severe abscessation and/or cellulitis and clostridial myonecrosis after IM injections.

References

[1] Williams PD. Intramuscular injection technique. Equine Vet Educ 1995;7:86–8.
[2] Ammann K. Die forensische Bedeutung der Injektionsschäden. Schweiz Arch Tierheilkd 1954;96:569–79.
[3] Zeller R. Über die haftpflicht des tierarztes bei injektionen. Prakt Tierarzt 1958;39:10–3.
[4] Koller A. Injektion und infektion [Dissertation]. Zürich: Veterinär-medizinische Fakultät,Universität Zürich; 1947.
[5] Deegen E. Intramuskuläre Injektionen und Komplikationen. 10 Bpt-Kongress. Münster: Vortragsband Kleintiere, Pferd; 2003. p. 100–1.
[6] Gerber H, Tercier P, Müller A. Injektionstechnik—injektionsfolgen beim pferd. Schweiz Arch Tierheilkd 1980;122:205–15.
[7] Ohnesorge B, Pfalzgraf S, Rohn K, Neuhaus M, Deegen E. Unverträglichkeitsreaktionen nach intramuskulärer injektion beim pferd—Auswertung einer Tierärztebefragung. Pferdeheilkunde 2006;22:337–46.
[8] Eikmeier H. Allgemeine sorgfaltspflichten des tierarztes. In: Eikmeier H, Fellmer E, Moegle H, editors. Lehrbuch der gerichtlichen tierheilkunde. 1st ed. Berlin und Hamburg: Verlag Paul Parey; 1990. p. 85–7.
[9] Eikmeier H. Injektionen. In: Eikmeier H, Fellmer E, Moegle H, editors. Lehrbuch der gerichtlichen Tierheilkunde. 1st ed. Berlin und Hamburg: Verlag Paul Parey; 1990. p. 88–95.
[10] Willener AW. Mechanik der injektion und verunreinigung des stichkanals [Dissertation]. Bern 1948.
[11] Heeg P, Bernau A, Dauber W. Infektionsgefahr durch injektionen. Biomed Eng 1990;35:179–81.
[12] Berge E, Westhues M. Tierärztliche operationslehre. 28th ed. Berlin: Paul Parey Verlag; 1961.
[13] Hutin Y, Hauri A, Chiarello L, Catlin M, Stilwell B, Ghebrehiwet T, et al. Best infection control practices for intradermal, subcutaneous, and intramuscular needle injections. Bull World Health Organ 2003; 81:491–500.
[14] Brown CM, Kaneene JB, Walker RD. Intramuscular injection techniques and the development of clostridial myositis or cellulitis in horses. J Am Vet Med Assoc 1988;193:668–70.
[15] Stanek C. Zur tiefen infiltration der langen rückenmuskulatur beim pferd: komplikationen und forensische aspekte. Pferdeheilkunde 2003;19:46–8.
[16] Rijkenhuizen A. Kann man ohne risiko intraartikulär injizieren? In: Akademie GmbH BPT, editor. 10 Bpt-Kongress, Münster. Vortragsband Kleintiere, Pferd; 2003. p. 102–4.

[17] Rose R, Hodgson D. Medication administration. In: Rose R, Hodgson D, editors. Manual of equine practice. 2nd ed. Philadelphia: W.B. Saunders Company; 2000. p. 752–6.

[18] Rossdale P, Ricketts S. Equine stud farm medicine. 2nd ed. London: Bailliere Tindall/Cassell Ltd; 1980.

[19] Koch G. Zur haftung des tierarztes bei injektionen unter besonderer berücksichtigung der desinfektion [Dissertation]. Berlin: Humboldt-Universität zu Berlin; 1966.

[20] Hanson DJ. Intramuscular injection injuries and complications. Am J Nurs 1963;63:99–101.

[21] Beecroft P, Redick S. Possible complications of intramuscular injections on the pediatric unit. Pediatr Nurs 1988;15:333–76.

[22] Garbade P. Veränderungen in der muskulatur nach intramuskulärer injektion beim pferd [Dissertation]. Hannover: Tierärztliche Hochschule; 1981.

[23] Eikmeier H. Stellungnahme zur arbeit "Erfahrungen mit der i.m. injektion in die seitliche halsmuskulatur beim pferd" von R. Tronicke Prakt Tierarzt 1976;57:220–4.

[24] Wissdorf H, Gerhards H, Huskamp B. Rumpfwand. In: Wissdorf H, Gerhards H, Huskamp B, Deegen E, editors. Praxisorientierte anatomie und propädeutik des pferdes. 2nd ed. Hannover: M. & G. Schaper GmbH; 2010. p. 315–63.

[25] Tronicke R. Erfahrungen mit der im injektion in die seitliche halsmuskulatur beim Pferd. Prakt Tierarzt 1976;57:12–4.

[26] Wissdorf H, Hertsch B, Geburek F, Fürst A, Huskamp B, Keller H. Oberschenkelbereich. In: Wissdorf H, Gerhards H, Huskamp B, Deegen E, editors. Praxisorientierte anatomie und propädeutik des pferdes. 2nd ed. Hannover: M. & H. Schaper GmbH; 2010. p. 591–7.

[27] Wissdorf H, Otto B, Gerhards H, Huskamp B. Hals. In: Wissdorf H, Gerhards H, Huskamp B, Deegen E, editors. Praxisorientierte anatomie und propädeutik des pferdes. 2nd ed. Hannover: M. & H. Schaper GmbH; 2010. p. 253–91.

[28] Boyd J. Selection of sites for intramuscular injections in the neck of the horse. Vet Rec 1987;121:197–200.

[29] Kollowa T. Die intramuskuläre injektion beim pferd. Prakt Tierarzt 1991;11:1003–5.

[30] Wissdorf H, Geburek F, Hertsch B, Fürst A, Huskamp B, Keller H. Beckenbereiche, kreuzdarmbeingelenk und hüftgelenk. In: Wissdorf H, Gerhards H, Huskamp B, Deegen E, editors. Praxisorientierte anatomie und propädeutik des pferdes. 2nd ed. Hannover: M. & H. Schaper; 2010. p. 598–617.

[31] Wissdorf H, Geburek F, Fürst A, Hertsch B, Keller H, Huskamp B. Bereich des M. triceps brachii. In: Wissdorf H, Gerhards H, Huskamp B, Deegen E, editors. Praxisorientierte anatomie und propädeutik des pferdes. 2nd ed. Hannover: M. & H. Schaper GmbH; 2010. p. 514–6.

[32] Gabka J. Komplikationen der intramuskulären injektion und ihre behandlung. In: Gabka J, editor. Die injektion; technik, praxis, komplikationen. 1st ed. Berlin: Walter de Gruyter & Co; 1968. p. 117–29.

[33] Eikmeier H. "Aseptische nekrosen" nach i.m. injektionen beim neugeborenen welpen. Berl Munch Tierarztl Wochenschr 1973;86:413–4.

[34] Marangi GF, Gigliofiorito P, Toto V, Langella M, Pallara T, Persichetti P. Three cases of embolia cutis medicamentosa (Nicolau's syndrome). J Dermatol 2010;37:488–92.

[35] Kienast AK, Mentze D, Hoeger PH. Nicolau's syndrome induced by intramuscular vaccinations in children: report of seven patients and review of the literature. Clin Exp Dermatol 2008;33:555–8.

[36] Brachtel R, Meinertz T. Local skin necroses after intramuscular injection—experimental animal studies. Arch Dermatol Res 1977;258:281–8.

[37] Müller-Vahl H, Pabst R. An animal model for aseptic necrosis after intramuscular injections. Int J Tissue React 1983;6:251–4.

[38] Stiehl P, Weissbach G, Schroter K. Das Nicolau syndrome. Schweiz Med Wochenschr 1971;99:266–9.

[39] Villarejo F, Pascual A. Injection injury of the sciatic nerve (370 cases). Childs Nerv Syst 1993;9:229–32.

[40] Gilles FH, French JH. Postinjection sciatic nerve palsies in infants and children. J Pediatr 1961;58:195–204.

[41] Combes MA, Clark WK, Gregory CF, James JA. Sciatic nerve injury in infants: recognition and prevention of impairment resulting from intragluteal injections. J Amer Med Assoc 1960;173:1336–9.

[42] Scheinberg L, Allensworth M. Sciatic neuropathy in infants related to antibiotic injections. Pediatrics 1957;19:261–5.

[43] Hochstetter A. Über probleme und technik der intraglutäalen injektion. Teil I. Schweiz Med Wochenschr 1955;85:1138–44.

[44] Kolb LC, Gray SJ. Peripheral neuritis as a complication of penicillin therapy. J Amer Med Assoc 1946;132:323–6.

[45] Hapke H-J. Störwirkungen bei tierarzneimitteln. Dtsch Tierarztl Wochenschr 1981;88:274–8.

[46] Hapke H-J. Unbeabsichtigte arzneimittelwirkungen. Tierarztl Umsch 1970;25:163–9.

[47] Eikmeier H. Gesundheitsschäden durch arzneimittel. Tierarztl Prax 1981;9:141–51.

[48] Pfalzgraf S. Unerwünschte arzneimittelwirkungen im zusammenhang mit der intramuskulären injektion beim pferd [Dissertation]. Hannover: Tierärztliche Hochschule Hannover; 2005.

[49] Oshida S, Degawa K, Takahashi Y, Akaishi S. Physico-chemical properties and local toxic effects of injectables. Tohoku J Exp Med 1979;127:301–16.

[50] Peek SF, Semrad SD, Perkins GA. Clostridial myonecrosis in horses (37 cases 1985-2000). Equine Vet J 2003;35:86–92.

[51] Valberg S, McKinnon A. Clostridial cellulitis in the horse: a report of five cases. Can Vet J 1984;25:67–71.

[52] Yuan J, Liu Y, Yang Z, Cai Y, Deng Z, Qin P, et al. Mycobacterium abscessus post-injection abscesses from extrinsic contamination of multiple-dose bottles of normal saline in a rural clinic. Int J Infect Dis 2009;13:537–42.

[53] Villanueva A, Villanueva Calderon R, Acosta Vargas B, Ruiz F, Aguero S, Zhang Y, et al. Report on an outbreak of postinjection abscesses due to mycobacterium abscessus, including management with surgery and clarithromycin therapy and comparison of strains by random amplified polymorphic DNA polymerase chain reaction. Clin Infect Dis 1997;24:1147–53.

[54] Rebhun W, Shin S, King J, Baum K, Patten V. Malignant edema in horses. J Am Vet Med Assoc 1985;187:732–6.

[55] Peek SF, Semrad SD. Clostridial myonecrosis in horses. Equine Vet Educ 2002;14:163–8.

[56] Breuhaus B, Brown C, Scott E, Ainsworth D, Taylor R. Clostridial muscle infections following intramuscular injections in the horse. J Equine Vet Sci 1983;3:42–6.

[57] MacKay R, Carlson G, Hirsh D. Clostridium perfringens associated with a focal abscess in a horse. J Am Vet Med Assoc 1979;175:71–2.

[58] Gehlen H, Venner M, Wohlsein P, Deegen E. Pararauschbrand bei einem pony nach intramuskulärer injektion. Pferdeheilkunde 1999;15:430–3.

[59] Mitten LA, Hinchcliff KW, Holcombe SJ, Reed SM. Mechanical ventilation and management of botulism secondary to an injection abscess in an adult horse. Equine Vet J 1994;26:420–3.

[60] Richey EJ. Clostridial (blackleg) diseases of cattle. University of Florida Cooperative Extension Service, Institute of Food and Agriculture Sciences, EDIS; 1996. p. 1–3.

[61] **Vengust M, Arroyo R, Weese JS, Staempfli H, Baird J. Clostridial myositis: evaluation of normal equine skeletal muscle for the presence of clostridial spores. 48th annual conference American Association of Equine Practitioners. Orlando, Florida; 2002. p. 134–135.**

[62] Deitch EA, Winterton J, Li M, Berg R. The gut as a portal of entry for bacteremia. Role of protein malnutrition. Ann Surg 1987;205:681–90.

[63] **Merck-Animal-Health.** http://www.merck-animal-health-usa.com/products/banamineinjectablesolution/overview.aspx, **[Accessed 19. 01.15].**

[64] Perdrizet J, Callihan D, Rebhun W, Shin S. Successful management of malignant edema caused by Clostridium septicum in a horse. Cornell Vet 1987;77:328–38.

[65] Anderson FL, Secombe CJ, Lester GD. Clostridial myonecrosis, haemolytic anaemia, hepatopathy, osteitis and transient hypertrophic cardiomyopathy after intramuscular injection in a Thoroughbred gelding. Aust Vet J 2013;91:204–8.

[66] Dyke TM, Sams RA, Cosgrove SB. Disposition of flunixin after intramuscular administration of flunixin meglumine to horses. J Vet Pharmacol Ther 1997;20:330–2.

[67] Pellegrini-Masini A, Poppenga R, Sweeney R. Disposition of flunixin meglumine injectable preparation administered orally to healthy horses. J Vet Pharmacol Ther 2004;27:183–6.

[68] Arata T, Kamitani M, Miyai T, Ito M. Antiseptic effects at injection sites. Dermatology 1997;195:107–10.

[69] Hague BA, Honnas CM, Simpson RB, Peloso JG. Evaluation of skin bacterial flora before and after aseptic preparation of clipped and nonclipped arthrocentesis sites in horses. Vet Surg 1997;26:121–5.

[70] Zubrod CJ, Farnsworth KD, Oaks JL. Evaluation of arthrocentesis site bacterial flora before and after 4 methods of preparation in horses with and without evidence of skin contamination. Vet Surg 2004;33:525–30.

[71] Svendsen O. Intramuscular injections and local muscle damage: an experimental study of the effect of injection speed. Acta Pharmacol Toxicol 1983;52:305–9.

[72] Kern O. Lokalverträglichkeit von arznei- und arzneihilfsstoffen bei intramuskulären injektionen Teil 3. Tierarztl Umsch 1987;42:971–2.
[73] Rasmussen F, Svendsen O. Tissue damage and concentration at the injection site after intramuscular injection of chemotherapeutics and vehicles in pigs. Res Vet Sci 1976;20:55–60.
[74] Mikaelian I, Poul J, Cabanie P. Healing of muscle trauma after intramuscular injection of antibiotics in sheep: correlations between clinical, macroscopic and microscopic scores. Vet Res 1996; 27:97–106.
[75] Lefebvre H, Laroute V, Braun J, Lassourd V, Toutain P. Non-invasive and quantitative evaluation of post-injection muscle damage by pharmacokinetic analysis of creatine kinase release. Vet Res 1996; 27:343–61.
[76] Steiness E, Rasmussen F, Svendsen O, Nielsen P. A comparative study of serum creatine phosphokinase (CPK) activity in rabbits, pigs and humans after intramuscular injection of local damaging drugs. Acta Pharmacol Toxicol (Copenh) 1978;42:357–64.
[77] Toutain PL, Lassourd V, Costes G, Alvinerie M, Bret L, Lefebvre HP, et al. A non-invasive and quantitative method for the study of tissue injury caused by intramuscular injection of drugs in horses. J Vet Pharmacol Ther 1995;18:226–35.
[78] Diness V. Local tissue damage after intramuscular injections in rabbits and pigs: quantitation by determination of creatine kinase activity at injection sites. Acta Pharmacol Toxicol (Copenh) 1985;56: 410–5.
[79] Gibson T, Norris W. Skin fragments removed by injection needles. Lancet 1958;2:983–5.
[80] Brandus V. The spinal needle as a carrier of foreign material. Can Anaesth Soc J 1968;15:197–201.
[81] Haindl H, Klinge O. Post injection abscess caused by non-suitable cannulas? Biomed Eng 1989;34:268–71.
[82] Klinge O, Müller H. Die bedeutung des kanülenschliffs für die pathogenese von spritzen-abszeß und-empyem. Dtsch Arztebl 1990;87:A-3742–6.
[83] Campbell DC, Douglas JM, Taylor G. Incidence of tissue coring with the 25-gauge Quincke and Whitacre spinal needles. Reg Anesth 1996;21:582–5.
[84] Piesold J, Müller W, Dreißig J. An experimental study on the aspirating reliability of different types of injection syringes with regard to the formation of punch cylinders. Br J Oral Maxillofac Surg 1998; 36:39–43.
[85] Puolakka R, Andersson LC, Rosenberg PH. Microscopic analysis of three different spinal needle tips after experimental subarachnoid puncture. Reg Anesth 2000;25:163–9.
[86] Baris S, Guldogus F, Baris YS, Karakaya D, Kelsaka E. Is tissue coring a real problem after caudal injection in children. Paediatr Anaesth 2004;14:755–8.
[87] Tunali Y, Kaya G, Tunali G, Solakoglu S, Yenice S, Bahar M. Detection of epithelial cell transfer in spinal areas by light microscopy and determining any tissue coring via cell culture during combined spinal-epidural interventions. Reg Anesth 2006;31: 539–45.
[88] Guldogus F, Baris YS, Baris S, Karakaya D, Kelsaka E. Comparing tissue coring potentials of hollow needles without stylet and caudal needles with stylet: an experimental study. Eur J Anaesthesiol 2008; 25:498–501.
[89] Belavy D. Brief reports: regional anesthesia needles can introduce ultrasound gel into tissues. Anesth Analg 2010;111:811–2.
[90] Bernau A, Cornelius C, Dauber W, Dietrich G, Heeg P. Hautstanzzylinder bei gelenkpunktionen. Orthop Praxis 1985;5: 359–65.
[91] Wahl K, Adams SB, Moore GE. Contamination of joints with tissue debris and hair after arthrocentesis: the effect of needle insertion angle, spinal needle gauge, and insertion of spinal needles with and without a stylet. Vet Surg 2012;41:391–8.
[92] Wagenknecht F. Beitrag zur technik der intravenösen injektion beim pferde [Dissertation]. Berlin: mechanische Vorbereitung des Operationsfeldes; 1924.

2.2 Publikation II:

Histological evaluation of equine cadaver skin cores dependent on needle type and skin preparation method to investigate intramuscular injection complications in horses

Published in: Journal of Equine Veterinary Science
Volume 45, October 2016, Pages 8–16

Received: June 15, 2015; Received in revised form: January 25, 2016; Accepted: February 4, 2016

Tobias Puschmann[1], Jan-Dirk Haeger[2], Nina Hambruch[2], Karl Rohn[3], Christiane Pfarrer[2†], Bernhard Ohnesorge[1†]

[1]Clinic for Horses, University of Veterinary Medicine Hannover, Foundation, Bünteweg 9, 30559 Hannover, Germany
[2]Institute of Anatomy, [3]Department of Biometry, Epidemiology and Information Processing, University of Veterinary Medicine Hannover, Foundation

[†] = equal contributors

Abstract:

Objective:
To histologically determine number and size of skin cores and hair fragments resulting from intramuscular injection techniques using equine cadaver skin explants.

Methods:
Clipped (*clip*) and non-clipped (*no.clip*) skin specimens, of the hamstring region (n = 10 horses), were obtained from freshly euthanized horses and punctured with 18G needles (18G), 18G needles with stylet (18GM), and 22G needles (22G). Needles were flushed with 0.2 ml of

sterile saline solution. The saline was collected on slides and stained (Pappenheim). Skin cores, produced with 18G, were fixed with formalin solution (4%) and embedded in paraffin. Sections were stained with Masson's trichrome. Light microscopy was used for evaluation.

Results:

Each different injection technique resulted in the identification of skin cores within a subset of slides obtained from the different 10 skin samples. Displaced skin cores were identified more frequently in *clip* samples than *no.clip* samples. 18G punctures resulted in the highest number and largest surface area of skin cores in both *no.clip* and *clip* samples as compared to 22G and 18GM punctures ($p<0.05$). 18GM punctures in *no.clip* samples resulted in the lowest number of skin cores.

Histology of paraffin sections identified connective tissue (14/16 samples), stratum corneum (13/16), epidermis (12/16), isolated hair shafts and sweat glands (11/16), hair shafts integrated with epidermis and sebaceous glands (6/16), and hair bulbs (5/16).

Conclusion:

Skin cores can contain bacteria which may initiate infective injection complications. The coring incidence as well as hair fragmentation correlates with needle size therefore smaller needle sizes should be used.

Journal of Equine Veterinary Science 45 (2016) 8–16

Contents lists available at ScienceDirect

Journal of Equine Veterinary Science

journal homepage: www.j-evs.com

Original Research

Histological Evaluation of Equine Cadaver Skin Cores Dependent on Needle Type and Skin Preparation Method to Investigate Intramuscular Injection Complications in Horses

Tobias Puschmann [a,*], Jan-Dirk Haeger [b], Nina Hambruch [b], Karl Rohn [c], Christiane Pfarrer [b,1], Bernhard Ohnesorge [a,1]

[a] Clinic for Horses, University of Veterinary Medicine Hannover, Foundation, Hannover, Germany
[b] Institute of Anatomy, University of Veterinary Medicine Hannover, Foundation, Hannover, Germany
[c] Department of Biometry, Epidemiology and Information Processing, University of Veterinary Medicine Hannover, Foundation, Hannover, Germany

ARTICLE INFO

Article history:
Received 15 June 2015
Received in revised form 25 January 2016
Accepted 4 February 2016
Available online 26 February 2016

Keywords:
Intramuscular injection
Complication
Skin core
Needle
Infection

ABSTRACT

To histologically determine number and size of skin cores and hair fragments resulting from intramuscular injection techniques using equine cadaver skin explants. Clipped (clip) and nonclipped (no.clip) skin specimens, of the hamstring region (n = 10 horses), were obtained from freshly euthanized horses and punctured with 18G needles (18G), 18G needles with stylet (18GM), and 22G needles (22G). Needles were flushed with 0.2 mL of sterile saline solution. The saline was collected on slides and stained (Pappenheim). Skin cores, produced with 18G, were fixed with formalin solution (4%) and embedded in paraffin. Sections were stained with Masson's trichrome. Light microscopy was used for evaluation. Each different injection technique resulted in the identification of skin cores within a subset of slides obtained from the different 10 skin samples. Displaced skin cores were identified more frequently in clip samples than no.clip samples; 18G needle punctures resulted in the highest number and largest surface area of skin cores in both no.clip and clip samples as compared to 22G and 18GM punctures ($P < .05$); 18G needles with stylet punctures in no.clip samples resulted in the lowest number of skin cores. Histology of paraffin sections identified connective tissue (14/16 samples), stratum corneum (13/16), epidermis (12/16), isolated hair shafts and sweat glands (11/16), hair shafts integrated with epidermis and sebaceous glands (6/16), and hair bulbs (5/16). Skin cores can contain bacteria which may initiate infective injection complications. The coring incidence and hair fragmentation correlate with needle size; therefore, smaller needle sizes should be used.

1. Introduction

Infective complications after injections include malignant edema and other inflammatory processes, which cumulate in cellulitis and abscess formation [1]. Cellulitis and abscesses both occur in 0.016% of annual intramuscular (IM) injections in horses [2]. There are two etiologies discussed in literature for injection site infections. Although it is suggested that bacteria gain access to injection sites via the blood stream, it is also suggested that with most injections, bacteria are directly advanced into deeper tissue layers by the needle either by displacement of skin cores or by translocation of microbes carried along the needle's surface [3–5]. The hypothesis of bacterial translocation was confirmed by isolating several bacteria species from the cultured needles after puncturing equine skin [6].

* Corresponding author at: Tobias Puschmann, Clinic for Horses, University of Veterinary Medicine Hannover, Bünteweg 9, 30559 Hannover, Germany.
E-mail address: tobias.puschmann@tiho-hannover.de (T. Puschmann).
[1] These authors contributed equally.

0737-0806/$ – see front matter © 2016 Elsevier Inc. All rights reserved.
http://dx.doi.org/10.1016/j.jevs.2016.02.231

Coring is a process also described by the pharmaceutical industry, defining it as particles cut out of the closure of drug vials, when puncturing it with needles. The skin itself is considered a semirigid material in which coring occurs [7]. Skin coring is described in numerous experimental studies in human medicine, comparing the coring potential of different needles with various types of skin specimens [8–18]. In these studies, the incidence of skin coring varies from 1.2% [13] to 97.1% [15], depending on the type of specimen, live or cadaver tissue from different body locations, and needles used. The types of cells and tissues identified by microscopic evaluation included epithelial cells [16,19], orthokeratotic keratin [17], adipose tissue [12], nucleated cells without mitotic activity [15], epithelial cell structures, and muscle fibers [14].

Skin cores occurring in lumbar punctures are discussed to be causative for the development of intraspinal epidermoid tumors in men [15–17]. Furthermore, skin cores occurring with intravenous punctures can be carried along the bloodstream and eventually reside in the lungs, in small arterioles or capillaries. Although these epithelial pulmonary emboli are likely removed after thrombus formation, it is suggested they introduce infective agents into the lung tissue [8]. The occurrence of hair and epidermal pulmonary emboli after intravenous treatments can be reported in mice [20], rats [21,22], rats and rabbits [23], dogs [24], and in one case report in men [25].

It is also suggested that subcutaneous or intramuscular skin cores in combination with a suitable microenvironment may result in superficial or deep postinjection infections such as abscesses or produce dermoid cysts [8,10] because bacteria may be advanced into deeper tissue layers adherent to skin cores [5,10]. Given the sterile production of pharmaceuticals, it is considered that skin cores be the primary carriers of pathogenic bacteria. Moreover, because injection sites are not sterile, the frequency of complications following IM injections may depend on the amount of translocated skin, hair fragments, and foreign debris [4]. However, given the relatively low frequency of reported postinjection infections [2], translocation of pathogenic bacteria adhered to skin cores may not necessarily produce infections in well perfused subcutis or muscle tissue. More likely, such pathogenic bacteria may initiate injection site infections in combination with administered pharmaceuticals, which cause tissue toxicity and thereby reduce local and systemic immune defense mechanisms [10]. Skin cores are suggested to carry a higher bacterial risk than hair fragments because they consist of sweat glands, sebaceous glands, and hair follicles [4]. Although few studies in humans including puncture of the thoracic cavity and disinfected knee joints isolated bacteria from skin cores [5,26,27], one study in horses did isolate bacteria from skin cores [28].

Few studies exist on skin coring in equids. An early study concerning intravenous punctures in horses indicated smaller sized needles had a lower coring potential and that coring occurred more frequently when injection sites were clipped, and most frequently when injection sites were shaved [29]. Another study identified hair fragments introduced into the injection canal via the needle insertion and that the highest incidence of hair fragment introduction was observed in the clipped injection sites [4].

Skin coring associated with equine arthrocentesis was studied by producing skin cores from skin sites including the fetlock joint capsule [28,30]. Tissue cores were observed with an incidence of 89% to 97% of punctures, whereas hair contamination was observed in 22.3% to 47.5%, depending on needle size, type, and preparation methods [30]. Wahl et al [28] published similar results, with a tissue coring incidence of 88.5%, hair contamination in 42%, and both hair and tissue contamination occurring in 39.2%. Hair fragments and hair follicles were embedded in tissue cores.

Considering the potential impact of skin coring on potential infective complications after needle punctures, and despite the existing human and equine research, the impact of cell and tissue transfer needs further investigation [12,19].

Therefore, one aim of this study was to compare the incidence, number, and size of the skin cores resulting from injections performed with different sizes and types of needles and different skin preparation protocols. Another aim was to determine the exact histologic components of skin cores so as to further investigate the impact of skin coring on IM complications in equids.

2. Materials and Methods

The skin specimens for this study were obtained from horses euthanized at the Clinic for Horses and the Department of Physiology of the University of Veterinary Medicine Hannover. All skin specimens were obtained within 2 hours of euthanasia. All horses were euthanized for reasons independent of this study (n = 10). Criteria for skin specimen donors included warmblood horses aged 1 year and older presenting with good general condition. Horses were not suffering from skin diseases or conditions such as systemic disease, endotoxemia, vasculitis/vasculopathy, or hemodynamic shock, and their hair coat was not secondarily contaminated or sweated. Sampling was performed with the euthanized horse in lateral recumbency. Square skin specimens, measuring 5 × 5 cm, were taken from the hamstring area, approximately 15 cm distal to the ischial tuberosity in the central region of the semimembranosis muscle. Peripherally, the sampling areas were clipped and cleaned with a 70% ethanol solution (Softasept N, B. Braun Melsungen AG, Germany), for minimizing contamination during sample acquisition. The skin was incised laterally and ventrally of the squares with both ventral corners fixed with surgical forceps. With the forceps kept under traction, the skin was separated from the underlying muscle tissue, allowing a clean sampling procedure. Eventually, the remaining dorsal tissue bridge was incised, providing the complete skin specimen. Each horse had skin samples obtained from two different skin preparations. Skin specimens type 1 (no.clip) were not clipped, whereas skin specimens type 2 (clip) were clipped. The skin specimens were placed on a sterilized test tube rack (Heinz Herenz Medizinalbedarf, Germany), between two platforms lying on top of each other and tightly fixed with screws (Fig. 1A–B). Each type of skin preparation was punctured with different needles including 18 gauge (G) needles without stylet (BD Microlance 3, Becton, Dickinson & Co. Ltd, Ireland), 18G needles with stylet (18GM) (BD Spinal needle, Becton Dickinson S.A., Spain), and 22G needles without stylet (BD Microlance 3, Becton Dickinson S.A., Spain). Skin punctures were created with subjectively extrapolated forces and speed needed for performing IM injections. To determine the number and size of skin cores obtained from different injection protocols, the needles were flushed using 0.2 ml of sterile saline solution. The effluent was immediately collected on glass slides, air-dried, stained with the Pappenheim's procedure and evaluated by light microscopy. Five slides representing five individual punctures from each of the six needle and skin preparation groups (i.e., a total of 50 slides generated from each puncture group) were produced from 10 horses for a total of 300 slides.

The number and surface area (measured in mm^2) of skin cores obtained per puncture (slide) were determined. The number of hair fragments per puncture was counted regardless of randomly occurring adjacent skin cores because with these methods, we were not able to differentiate between hair fragments which derive from the skin core itself or lose ones being coincidentally situated with the skin cores. Skin cores and hair fragments were analyzed separately. Microscopic software (Leica Application Suite Mikroskop-Software, Leica, Germany) was used for evaluation.

For production of paraffin-embedded sections, three additional skin cores from each horse were obtained from the no.clip specimens only, using 18G needles (BE Microlance 3, Becton, Dicinson & CO. Ltd, Ireland). To enhance visibility of extremely small skin core samples for further histoprocessing and paraffin wax embedding, after fixation with formalin solution (4%), skin cores were embedded in agarose (3%) similarly to descriptions in the literature [31–33]. The agarose was poured on the skin cores which were placed in test tubes. After solidification, the agarose block was extracted and trimmed into a cubic shape for paraffin embedding, which was performed as previously described in literature [34]. Masson staining was performed using a standard protocol described previously in literature [35]. Evaluation was performed using light microscopy (Laborlux, Leica, Germany) with magnifications ranging from 6.3 to 40 (objective) × 12.5 (ocular).

2.1. Statistical Analysis

Statistical analysis was performed using statistic software SAS, version 9.3 (SAS Institute, Cary, NC), considering P-values $< .05$ significant. Mean values were calculated for the total number of skin cores and hair fragments, and the surface areas (in mm^2) of all skin cores obtained from each slide representing each sample group. Any of the five slides of a sample group, which did not contain skin cores or hair fragments, was excluded from statistical analysis of the number of hair fragments, number of skin cores, and surface areas because these zero values would have wrongly lowered the actual mean values of these parameters. For the same reason, the whole sample group was excluded if all five slides did not contain skin cores or hair fragments. Therefore, different numbers of the sample groups were included in the statistical analysis. Mean values of each sample group were used for further statistic evaluation. The assumption of normality was tested using the Kolmogorov–Smirnov test and visual assessment of qq-plots of model residuals. Because of rejection of normal distribution for all data, distribution-free nonparametric methods were applied. One-way analysis with repeated measurements was used for comparing the quantity and size of skin cores produced, within each preparation protocol, depending on needle size and type. For that permutation test with post hoc multiple pairwise comparisons and Sidak adjustment for P-values was calculated. For the comparison of quantity and size of skin cores between skin specimen types, depending on needle type and size, two sample signed rank test was used. The incidence of skin fragmentation was evaluated using Fischer's exact test. The incidence of hair fragmentation was evaluated using chi-square testing and Fischer's exact test when numbers were too low for chi-square testing. Paraffin sections were evaluated descriptively.

3. Results

For each sample group comprising 50 slides representing each separate puncture made through each different skin preparation (clip vs. no.clip) by each different needle type and size (18G, 18GM, and 22G), overall lower numbers of skin cores were obtained from the unclipped preparations than the clipped preparations with the lowest numbers of skin cores obtained with the 18GM needles

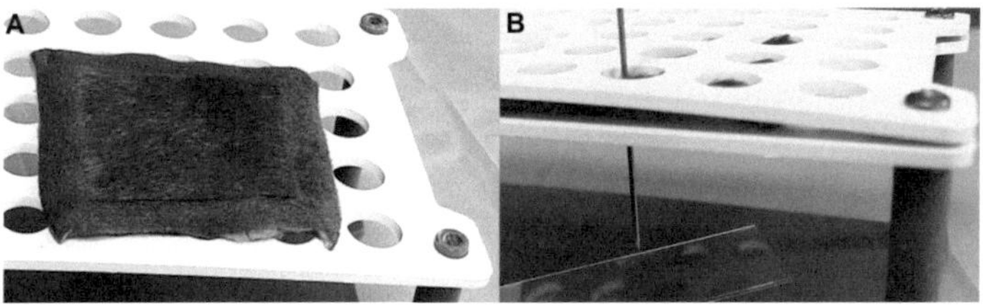

Fig. 1. (A) Skin specimen (no.clip) placed on the first platform of a sterilized test tube rack. (B) Skin specimen (no.clip), punctured with an 18G needle while being tightly fixed with screws between two platforms of a test tube rack. A slide is positioned underneath the needle's tip to collect the saline solution used for flushing skin fragments out of the needle's lumen.

Table 1
Incidence of hair fragmentation and number of hair fragments per puncture.

Needle Type and Size and Skin Specimen Type	Number of Puncture Groups Included in Statistical Analysis (n = 10)	Number (With Percentage) of Slides/Punctures With Hair Fragments (n = 50) (%)	Number of Hair Fragments Per Slide/ Puncture (n =#)		
			Median	Mean	Range
18G no.clip	3/10	7/50 (14)	3	3.43	1–8
18GM no.clip	8/10	11/50 (22)	1	1.36	1–3
22G no.clip	3/10	3/50 (6)*	1	1	1–1
18G clip	6/10	12/50 (24)	1.5	1.92	1–4
18GM clip	10/10	20/50 (40)	2	2.25	1–6
22G clip	5/10	6/50 (12)**	1	1	1–1

Abbreviations: 18G, 18G needles; 18GM, 18G needles with stylet; 22G, 22G needles.
= only slides on which hair fragments were detected were used for median, mean, and range calculations, of n = 50 slides; * = significant to 18GM no.clip (*P*-value < .05); ** = very significant to 18GM clip (*P*-value < .01); all other comparisons were nonsignificant.

(28/50, 56%) puncture performed on the no.clip preparation. Compared to 18G needles (18G) with no.clip preparations, there was a significant incidence decrease observed when 22G (*P*-value < .01) and 18GM needles (*P*-value < .001) were used. Skin coring per puncture was highest with 18G for both clip and no.clip preparations (48/50, 96% and 46/50, 92%, respectively). The skin coring incidence increased in 18GM significantly (no.clip: 28/50, 56% to clip: 43/50, 86%; *P*-value < .01) and nonsignificantly in 22G needle (22G)(no.clip: 36/50, 72% to clip: 42/50, 84%) punctures when clip specimens were used instead of no.clip specimens. The comparison of the skin coring incidence of 18G between clip and no.clip turned out nonsignificant, as well as comparison of skin coring incidences between all needle types with clip preparations, and incidences between 18GM and 22G in no.clip preparations (Fig. 2). Overall, hair fragments were observed more frequently when punctures were performed using clipped skin specimens with all needle types and sizes. The comparison of the incidence of hair fragmentation between clip and no.clip for each needle type turned out nonsignificant. In both clip and no.clip skin specimens, 18GM needles produced the highest number of hair fragments, whereas 22G punctures produced the lowest number of hair fragments (no.clip: *P*-value < .05; clip: *P*-value < .01) (Table 1). The quantity and incidence of skin coring and hair fragmentation varied substantially between several specimens when comparing the same specimen type with the usage of the same needle type. Hence, with 18GM needles and no.clip specimens, only nine skin specimens were included in statistical testing because with one skin specimen, no skin coring was observed in the whole sample group, whereas with others, skin coring occurred in all five slides of a sample group.

The quantity of skin cores per puncture obtained from puncturing no.clip and clip skin specimens was reduced significantly when 18GM and 22G were used instead of 18G. No significant difference of skin core quantity was observed when skin preparation types were compared (Fig. 3).

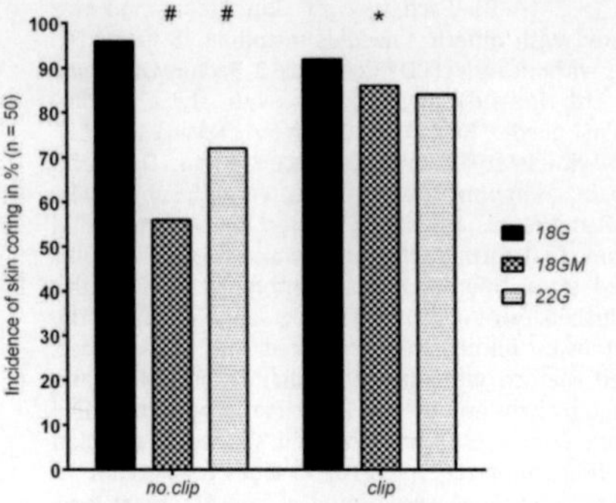

Fig. 2. Column diagram showing the skin coring incidence of 18G, 18GM, and 22G needles grouped by no.clip and clip preparation protocols. # = significant to 18G no.clip (18GM *P*-value < .001; 22G *P*-value < .01); * = significant to 18GM no.clip (*P*-value < .01); all other comparisons were nonsignificant. 18G, 18G needles; 18GM, 18G needles with stylet; 22G, 22G needles.

Considering the skin core size, punctures with 18G produced significantly larger skin fragments than 18GM needles and 22G, respectively. However, also with skin core size, there was no significant difference observed between no.clip and clip specimens for each needle type (Fig. 4).

Of the 30 samples that underwent paraffin sectioning, 16 were evaluable. Connective tissue was observed most frequently in 14/16 (87.5%) samples, followed by stratum corneum in 13/16 (81.2%) samples, full-thickness epidermis in 12/16 (75%) samples, isolated hair shafts and sweat glands in 11/16 (68.7%) samples, hair shafts integrated with epidermis and sebaceous glands in 6/16 (37.5%) samples, and hair bulbs in 5/16 (31.2%) samples (Figs. 5–6).

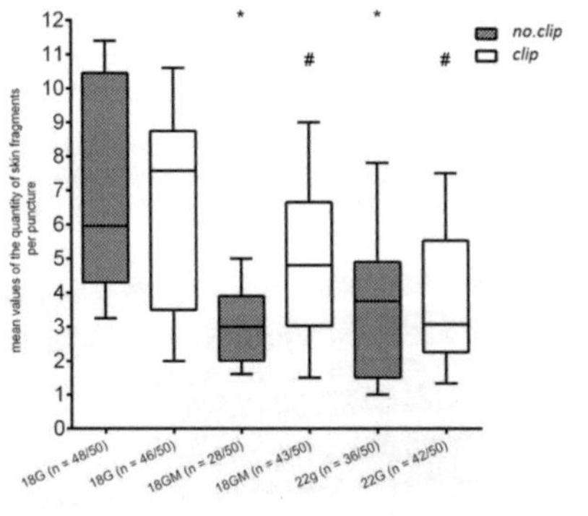

Fig. 3. Box-and-whisker plot (Tukey method) showing the comparison of the mean values of the quantity of skin fragments per puncture, grouped by needle type and size and skin specimen type. * = highly significant to 18G no.clip (*P*-value < .001); # = significant to 18G clip (18GM *P*-value < .05; 22G *P*-value < .001); all other comparisons were nonsignificant. 18G, 18G needles; 18GM, 18G needles with stylet; 22G, 22G needles.

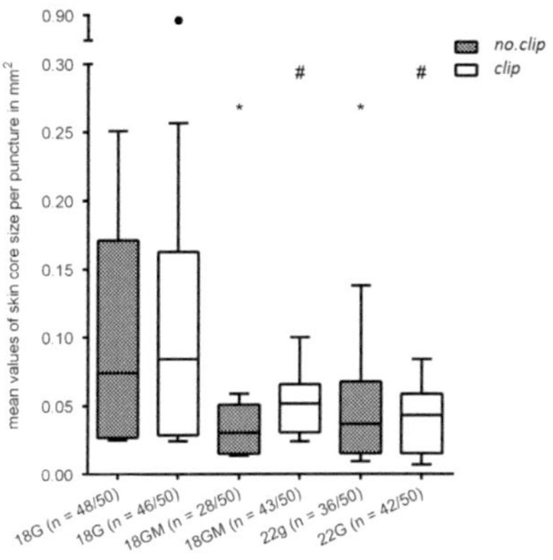

Fig. 4. Box-and-whisker plot (Tukey method) showing the comparison of the mean values of skin core sizes grouped by needle type and size and skin specimen type. * = very significant to 18G no.clip (*P*-value < .01); # = significant to 18G clip (*P*-value < .01); ● = outlier; all other comparisons were nonsignificant. 18G, 18G needles.

4. Discussion and Conclusion

The criteria for horses to qualify as skin specimen donors were chosen to allow a good comparability of the results. Furthermore, criteria included only skin specimens from the hamstring area, as the composition of equine skin was found to vary between body locations [36]. The quantity and incidence of skin coring and hair fragmentation varied comparing the same specimen type with the usage of the same needle type. Similarly to this observation, Wahl et al [28] describe an increased odds ratio of hair fragmentation, when using identical needle and specimen types, as described in a preceding study performed by Adams et al [30]. We cannot explain the variation of the quantity and incidence of skin coring or hair fragmentation in this study. Possibly even within the warmblood breeds, individuals differ in skin and hair constitution, leading to the observed difference of skin coring and hair fragmentation incidence. Furthermore, the puncturing process was not standardized in our protocol, as it was described by Piesold et al, who were using a mechanical puncturing device [13]. Therefore, slight deviations in speed, force, and insertion angle, causing variations in the coring incidence, must be expected. But as injections in equine practice are not performed by mechanical devices, this study design may, nevertheless, produce findings similar to their occurrence in the field, when IM injections are performed in living horses. Because the preceding unpublished study used to establish this study's protocol revealed large variations of the coring incidence, it was decided to produce five samples from each individual skin preparation and needle variation so as to calculate the mean values of each sample group for further evaluation. Only positive slides from each representative sample group were used to calculate mean values of the dimensions and other parameters of skin cores. We chose the hamstring area for skin harvesting because it is an appropriate location where IM injections are preformed regularly [37–39]. Because in the literature recommended sizes of needles for IM injections in horses include 18G to 21G needles [39,40], we decided to compare 18G and 22G. The comparison between clipped and unclipped skin specimens was chosen because clipping before IM injections is advised if the hair is long [37]. Furthermore, in some studies, the reported skin coring and hair fragmentation incidences are increased in clipped skin preparations [4,28–30]. Using a test tube rack as a sampling device, although not yet described in the literature, proved to be a practical and clean procedure, which minimized secondary contaminations of the skin specimens, and allowed for fixation of skin samples.

Although the possibility exists that the 0.2 ml amount of saline used to flush the needles to recover all intraluminal particulates (i.e., skin cores and hair fragments) was insufficient, previous unpublished experiments performed by the authors showed that the number of macroscopically visible particulates did not increase with increasing volumes of saline. Compared to other simple staining procedures that did not require prefixation, the Pappenheim stain, a procedure generally used in hematology [41], provided the best staining to differentiate skin cores from other debris and therefore was selected for this study.

We consider the results of the numerous studies on skin coring from human medicine only partly comparable, as animal skin with its hair coat differs from human skin. Because the equine studies showing the closest resemblance to our study were performed by Adams et al [30] and Wahl et al [28] using skin specimens including the joint capsules, we chose to compare our results primarily with these two studies.

Our results confirm that coring occurs more frequently with larger needle diameters, which coincides with findings of other studies [4,28,30]. The largest needle size used in our study was an 18G, which led to a higher incidence of coring in no.clip specimens, compared to 22G. This could be expected, as larger needle lumina increase the risk of skin cores being produced, or debris being advanced through the skin. Compared to 18G, 18GM needles show a lower coring incidence if the skin is not clipped. This finding is comprehensible, as hollow needles must have a higher coring potential than needles with stylet, which were developed to reduce skin coring. Nevertheless, these findings do not coincide with the results published by Wahl et al [28] and Adams et al [30]. In both studies, using skin specimens including joint capsules, 20G hollow needles produced fewer skin cores than 20G spinal needles.

In our study, clipping increased the incidence of coring with both 18GM and 22G, whereas it was slightly decreased with 18G. Clipping the hair possibly makes it easier for needles to cut sharply into the skin and curve out tissue fragments, which may explain the increase of the coring incidence with 22G and 18GM needles. Willener, as well as Wagenknecht, report that coring increases when the hair is clipped compared to unclipped hair [4,29]. Contrarily, Wahl et al [28] published results indicating a decrease of the coring incidence when the hair is clipped. We only

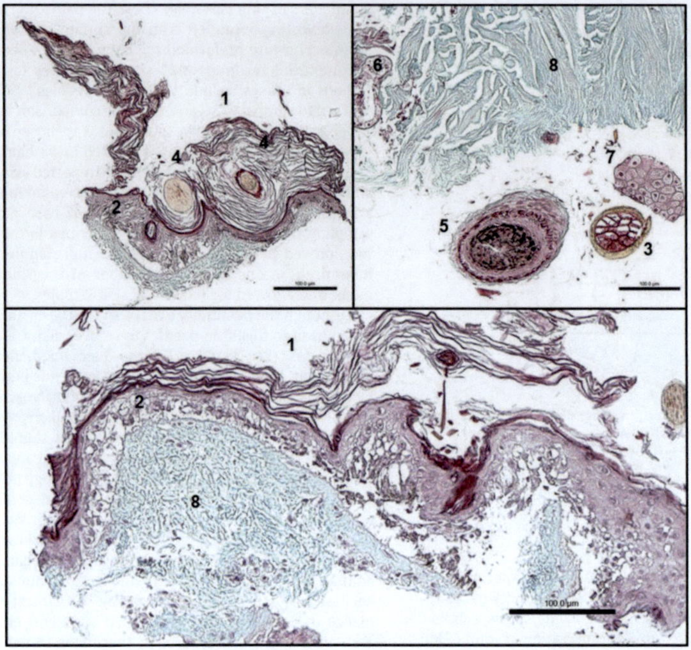

Fig. 5. Representative masson-stained paraffin sections of skin cores. 1 = stratum corneum of epidermis; 2 = vital layers of epidermis; 3 = isolated hair shaft; 4 = hair integrated in epidermis; 5 = hair bulb; 6 = sweat gland; 7 = sebaceous gland; and 8 = dermal connective tissue. Given the short interval from euthanasia to tissue harvest, the epidermal clearing and dermoepidermal clefting are likely artifacts associated with tissue procurement.

observed a slight decrease in the coring incidence with 18G when the hair was clipped, which might be coincidental and might change with higher sample numbers.

Considering the quantity of skin cores per puncture, a reduced number significantly correlated to needle size but not type of skin preparation (e.g., clipped vs. unclipped).

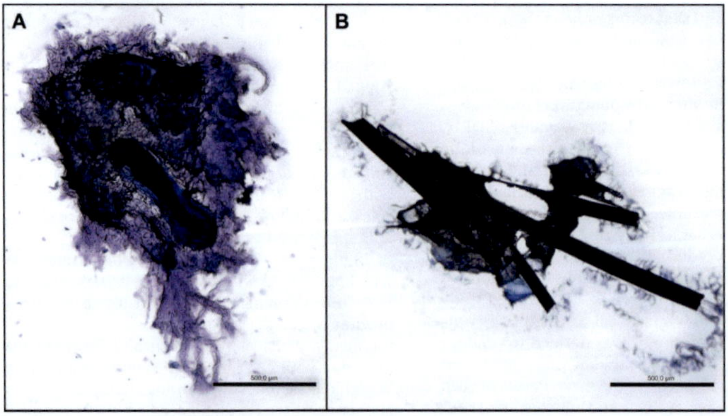

Fig. 6. Pappenheim stained skin cores, as used for size, quantity, and hair fragment evaluation. (A) skin core with hair bulb; (B) skin core with hair fragments adjacent to each other.

This indicates that the overall incidence of coring and the quantity of skin cores per puncture correlates with needle size and type, not whether the skin was clipped or unclipped. However, the quantity of skin cores per puncture does not appear to be the most representative parameter for risk evaluation. Rather, we consider the total size of all skin cores produced by one puncture a more specific parameter for risk evaluation because larger skin cores may advance larger amounts of bacteria containing structures through the skin, which, depending on type and number of bacteria, possibly result in a higher risk for infection.

Although the total size of skin cores per puncture was significantly reduced in both clipped and unclipped specimen types when using the smaller 22G and 18GM, the overall number of hair fragments per skin puncture was highest with the 18GM needles in clipped skin. Interestingly, the 18-gauge needles with a stylet in unclipped skin produced a near equal number of hair fragments per puncture as compared to hollow 18-gauge needles in clipped skin. The incidence of hair fragmentation was highest with 18GM needles, followed by 18G and 22G in both specimen types, with significant differences between 18GM needles and 22G in both specimen types. The expected lower incidence of hair fragmentation using 22G versus 18G in our study correlates with needle size, as previously described by Adams et al [30], who observed increasing amounts of hair fragments when bigger needle sizes were used. Furthermore, our results coincide with other reports in the literature describing an increased number of hair fragments transferred through the skin by needles with stylets (e.g., 18GM) in comparison to hollow needles without stylets [28,30]. A possible mechanism proposed in both studies includes an imperfect fit of the stylet within the needle shaft that results in trapping hairs. Wahl et al [28] also reported a higher incidence of hair fragment transfer that occurred with higher gauge spinal needles inserted without their stylets in place and proposed this resulted from the relatively shorter and less oblique tips of spinal needles as compared to standard hollow needles. However, comparison of clipped versus unclipped skin specimens did not reveal any significant differences in the number of hair fragments produced per puncture between 18G, 18GM, or 22G.

The coring incidence, skin core size, and quantity, as well as the event of hair fragmentation, show tendencies of increase in this study in most punctures, if the skin is clipped before needle insertions. Therefore, clipping before performing IM injections cannot be advised for these reasons. It must be considered, nevertheless, to clip horse skin before injections, if the hair coat is soiled, as in those cases the bacteriological risk may exceed the potential risk presented by an increase of skin cores and hair fragments.

Paraffin sectioning turned out to be a valuable procedure for an exact analysis of the histologic components of the skin cores. Because of practical and financial limitations of this study, we were only able to produce paraffin sections of one specimen type. We chose no.clip specimens because IM injections are usually performed without clipping the hair. To identify and differentiate the various skin components (e.g., skin and adnexa from dermal connective tissue), a Masson's trichrome stain was selected for the paraffin-embedded skin cores. However, using this method would be desirable for future studies, for example to investigate possible morphologic differences histologically between skin cores obtained from differently prepared skin specimens. The embedding procedure was performed as described previously by Korff and Augustin [34]. However, because only 16 of 30 sample were evaluable, this method is not suitable for determination of actual skin coring incidence and average skin core size associated with a particular puncture technique for the following reasons. Unfortunately, positioning of some skin cores at the edge of the agarose blocks will result in excision of these skin cores during trimming of the solidified agarose into square blocks before paraffin embedding. In addition, skin cores that were not macroscopically visible precluded proper positioning in the agarose, thus resulted in a loss of these microscopic samples. In addition, some skin cores were curled up on themselves, making evaluation of true size impossible. Thus, average incidence of skin cores might be falsely underestimated, whereas average size of the skin cores might be falsely overestimated (if microscopic samples are lost) or underestimated (if skin cores are curled). Utilization of a dissecting microscope may prevent this loss from occurring and might also allow for uncurling and proper positioning of samples [33].

However, the paraffin embedding with Masson's histochemical stain did provide good identification of skin and adnexal structures included in skin cores (as presented in Figs. 4 and 5) and allowed the authors to present the first description of skin cores associated with equine cutaneous needle punctures.

Hair follicles [42], the different layers of the stratum corneum [42,43], pilosebaceous follicles [44], and sweat ducts [45] are structures in which skin bacteria were detected in human medicine. Other studies have confirmed epithelial cells [16,19], nonnucleated stratum corneum cells [17], nucleated cells [15], fat tissue [12], epithelial cell structures and muscle fibers [14], tissue and hair fragments, hair follicles, and hair embedded in tissue clumps [28,30]. Skin cores in this study contained variable components of the skin; among these were connective tissue, stratified stratum corneum, epidermis, isolated hair shafts, sweat glands, hair shafts integrated with epidermis and sebaceous glands, and hair bulbs. The different incidences with which individual structures were detected may most likely a coincidental finding. The changes to the epidermis, such as dermoepidermal clefting, intracellular edema of the spinous layers, are more probably artifacts related to the sampling procedure than to autolysis because skin specimens were harvested immediately after euthanasia.

We conclude from the results of our study that skin cores must be anticipated to contain bacteria inhabited structures of the skin. Therefore, their role as an infective agent must be discussed. It is unclear if the bacterial load can be the sole initiator of infective injection complications, causing abscess formation or cellulitis, or if tissue alterations at the injection site resulting from slightly irritating drugs reduce local immunity and/or promote bacterial growth. In addition, subcutaneous displacement of epidermis and adnexal structures alone could produce sterile inflammation through a foreign body reaction. Our assumption that subcutaneous skin core displacement can result in postinjection infections is supported by the results of several studies reporting positive bacterial growth after culturing needles' content from aseptically prepared skin punctures [5,27,28]. Our results indicate that the incidence of skin coring, the quantity of skin cores per puncture, the summarized total size of skin cores per puncture, and the quantity of hair fragments increase when larger needle sizes are used. Given results from this and other studies, because hair clipping under certain conditions is associated with an increased incidence in hair fragments and skin cores, if not needed for other reasons (e.g., over long and soiled haircoat), the authors recommend against it. The use of needles with stylets is disadvantageous for IM injections because they lead to a higher incidence of hair fragmentation. This has to be considered specifically before performing epidural injections, for which injection sites are regularly clipped, and thus an increased coring incidence should be expected. In conclusion, skin cores (may) contain bacteria inhabited skin structures which may initiate infective injection complications. The coring incidence and hair fragmentation correlate with needle size; therefore, smaller needle sizes should be used.

Acknowledgments

The authors acknowledge the contributions of Doris Walter and Marion Gähle for the support with sample processing procedures at the Department of Histology of the Institute of Anatomy, University of Veterinary Medicine Hannover, Dr. Klaus Hopster DipECVAA for writing assistance and all coworkers for assistance with sample acquisition at the Clinic for Horses, University of Veterinary Medicine Hannover. All authors disclose that there are no conflicts of interest which inappropriately influence/bias this work.

References

[1] Koch G. Zur Haftung des Tierarztes bei Injektionen unter besonderer Berücksichtigung der Desinfektion [Dissertation]. Berlin: Humboldt-Universität zu Berlin; 1966.
[2] Ohnesorge B, Pfalzgraf S, Rohn K, Neuhaus M, Deegen E. Unverträglichkeitsreaktionen nach intramuskulärer Injektion beim Pferd - Auswertung einer Tierärztebefragung. Pferdeheilkunde 2006;22:337–46.
[3] Ammann K. Die forensische Bedeutung der Injektionsschäden. Schweiz Arch Tierheikd 1954;96:569–79.
[4] Willener AW. Mechanik der Injektion und Verunreinigung des Stichkanals [Dissertation]; 1948. Bern.
[5] Heeg P, Bernau A, Dauber W. Infektionsgefahr durch Injektionen. Biomed Eng 1990;35:179–81.
[6] Koller A. Injektion und Infektion [Dissertation]. Zürich: Veterinärmedizinische Fakultät, Universität Zürich; 1947.
[7] Charlebois PA. Coring: the unseen menace. Can Anaesth Soc J 1966; 13:585–97.
[8] Gibson T, Norris W. Skin fragments removed by injection needles. Lancet 1958;2:983–5.
[9] Brandus V. The spinal needle as a carrier of foreign material. Can Anaesth Soc J 1968;15:197–201.
[10] Haindl H, Klinge O. Post injection abscess caused by non-suitable cannulas? Biomed Eng 1989;34:268–71.
[11] Klinge O, Müller H. Die Bedeutung des Kanülenschliffs für die Pathogenese von Spritzen-Abszeß und-Empyem. Dtsch Arztebl 1990;87:A-3742–6.
[12] Campbell DC, Douglas JM, Taylor G. Incidence of tissue coring with the 25-gauge Quincke and Whitacre spinal needles. Reg Anesth 1996;21:582–5.
[13] Piesold J, Müller W, Dreißig J. An experimental study on the aspirating reliability of different types of injection syringes with regard to the formation of punch cylinders. Br J Oral Maxillofac Surg 1998; 36:39–43.
[14] Puolakka R, Andersson LC, Rosenberg PH. Microscopic analysis of three different spinal needle tips after experimental subarachnoid puncture. Reg Anesth Pain Med 2000;25:163–9.
[15] Baris S, Guldogus F, Baris YS, Karakaya D, Kelsaka E. Is tissue coring a real problem after caudal injection in children. Pediatr Anesth 2004; 14:755–8.
[16] Tunali Y, Kaya G, Tunali G, Solakoglu S, Yenice S, Bahar M. Detection of epithelial cell transfer in spinal areas by light microscopy and determining any tissue coring via cell culture during combined spinal-epidural interventions. Reg Anesth Pain Med 2006;31:539–45.
[17] Guldogus F, Baris YS, Baris S, Karakaya D, Kelsaka E. Comparing tissue coring potentials of hollow needles without stylet and caudal needles with stylet: an experimental study. Eur J Anaesthesiol 2008; 25:498–501.
[18] Belavy D. Brief reports: regional anesthesia needles can introduce ultrasound gel into tissues. Anesth Analg 2010;111:811–2.
[19] Özyurt G, Moğol E, Tolunay S, Kerimoğlu B. Tissue coring with spinal needles. Reg Anesth Pain Med 2000;25:665.
[20] Innes JR, Donati EJ, Yevich PP. Pulmonary lesions in mice due to fragments of hair, epidermis and extraneous matter accidentally injected in toxicity experiments. Am J Pathol 1958;34: 161–7.
[21] Tekeli S. Occurrence of hair-fragment emboli in the pulmonary vascular system of rats. Vet Pathol 1974;11:482–5.
[22] Morton D, Safron JA, Glosson J, Rice DW, Wilson DM, White RD. Histologic lesions associated with intravenous infusions of large volumes of isotonic saline solution in rats for 30 days. Toxicol Pathol 1997;25:390–4.
[23] Kast A, Tsunenari Y. Hair embolism in lungs of rat and rabbit caused by intravenous injection. Lab Anim 1983;17:203–7.
[24] Schneider P, Pappritz G. Hairs causing pulmonary emboli. A rare complication in long-term intravenous studies in dogs. Vet Pathol 1976;13:394–400.
[25] Andrew JH. Pulmonary skin embolism: a case report. Pathology 1976;8:185–7.
[26] Little DC. Skin fragments in endopening needles. Can Med Assoc J 1955;72:374–5.
[27] Bernau A, Cornelius C, Dauber W, Dietrich G, Heeg P. Hautstanzzylinder bei Gelenkpunktionen. Orthop Praxis 1985;5: 359–65.
[28] Wahl K, Adams SB, Moore GE. Contamination of joints with tissue debris and hair after arthrocentesis: the effect of needle insertion angle, spinal needle gauge, and insertion of spinal needles with and without a stylet. Vet Surg 2012;41:391–8.
[29] Wagenknecht F. Beitrag zur Technik der intravenösen Injektion beim Pferde: mechanische Vorbereitung des Operationsfeldes [Dissertation]; 1924. Berlin.
[30] Adams SB, Moore GE, Elrashidy M, Mohamed A, Snyder PW. Effect of needle size and type, reuse of Needles, insertion speed, and removal of hair on contamination of joints with tissue debris and hair after arthrocentesis. Vet Surg 2010;39:667–73.
[31] Arnolds W. Oriented embedding of small objects in agar-paraffin, with reference marks for serial section reconstruction. Stain Technol 1978;53:287–8.
[32] Van der Wal U, Dohmen M. A method for the orientation of small and delicate objects in embedding media for light and electron microscopy. Stain Technol 1978;53:56–8.
[33] Blewitt E, Pogmore T, Talbot I. Double embedding in agar/paraffin wax as an aid to orientation of mucosal biopsies. J Clin Pathol 1982; 35:365.
[34] Korff T, Augustin HG. Integration of endothelial cells in multicellular spheroids prevents apoptosis and induces differentiation. J Cell Biol 1998;143:1341–52.
[35] Bancroft JD, Gamble M. Theory and practice of histological techniques. 8th ed. Philadelphia, USA: Churchill Livingstone Elsevier; 2008.
[36] Monteiro-Riviere NA, Bristol DG, Manning TO, Rogers RA, Riviere JE. Interspecies and interregional analysis of the comparative histologic thickness and laser Doppler blood flow measurements at five cutaneous sites in nine species. J Invest Dermatol 1990;95:582–6.

[37] Deegen E. Intramuskuläre Injektionen und Komplikationen. 10 Bpt-Kongress, Münster. 2003; Vortragsband Kleintiere, Pferd: p.100–101.
[38] Garbade P. Veränderungen in der Muskulatur nach intramuskulärer Injektion beim Pferd [Dissertation]. Hannover: Tierärztliche Hochschule; 1981.
[39] Rose R, Hodgson D. Medication administration. In: Rose R, Hodgson D, editors. Manual of equine practice. 2nd ed. Philadelphia: W.B. Saunders Company; 2000. p. 752–6.
[40] Rossdale P, Ricketts S. Equine stud farm medicine. 2nd ed. London: Bailliere Tindall/Cassell Ltd.; 1980.
[41] Begemann H, Rastetter J. Staining methods. Atlas of clinical haematology. Germany: Springer Berlin Heidelberg; 1972. p. 9–21.
[42] Lange-Asschenfeldt B, Marenbach D, Lang C, Patzelt A, Ulrich M, Maltusch A, et al. Distribution of bacteria in the epidermal layers and hair follicles of the human skin. Skin Pharmacol Physiol 2011; 24:305–11.
[43] Montes LF, Wilborn WH. Location of bacterial skin flora. Br J Dermatol 1969;81:23–6.
[44] Puhvel SM, Reisner RM, Amirian DA. Quantification of bacteria in isolated pilosebaceous follicles in normal skin. J Invest Dermatol 1975;65:525–31.
[45] Malcolm SA, Hughes TC. The demonstration of bacteria on and within the stratum corneum using scanning electron microscopy. Br J Dermatol 1980;102:267–75.

2.3 Publikation III:

Evaluation of bacterial contamination of equine cadaver skin following injections using different needle types and skin preparation methods.

Published in: Journal of Equine Veterinary Science
Volume 46, November 2016, Pages 24–30

Received: February 8, 2016; Received in revised form: June 28, 2016; Accepted: June 28, 2016

T. Puschmann[1], J. Verspohl[2], K. Rohn[3], B. Ohnesorge[1]
[1]Clinic for Horses, [2]Institute for Microbiology, [3]Department of Biometry, Epidemiology and Information Processing, University of Veterinary Medicine Hannover, Foundation

Abstract:

Objective:
To assess bacterial translocation through needle punctures and the impact of preparation protocols using equine cadaver skin and different needle sizes and types.

Methods:
Non-prepared, disinfected, clipped and disinfected, and clipped and antiseptically washed skin specimens were sampled sterilely using swabs and biopsies to compare disinfection efficacy. The skin specimens were punctured using 18G hollow needles (18G), 18G needles with stylet (18GM) and 22G hollow needles (22G), and were flushed with sterile saline solution. The saline was collected in blood culture systems or homogenization test tubes and processed for qualitative (ql) or semiquantitative microbiological evaluation.

Results:

Disinfection and antiseptic preparation of skin specimens led to a significant reduction of bacteria ($p < 0.05$) sampled with swabs and biopsies, whereas clipping did not produce further significant differences. Bacteria were detected in many needle punctures, depending on the preparation protocols and needle types and sizes used. The highest amount of bacteria was detected with 18GM, and the lowest with 22G (ql; non-prepared; nonsignificant). Disinfection and antiseptic preparation reduced bacterial translocation significantly in 18G punctures (ql; $p < 0.05$), whereas clipping did not result in any further significant difference. Overall, more bacteria were detected after 18G needle punctures when samples were processed qualitatively.

Conclusion:

Disinfection prior to injections reduces the number of bacteria translocated into deeper tissue. Larger needles and needles with stylet increase the number of bacteria translocated through the skin.

Journal of Equine Veterinary Science 46 (2016) 24–30

Contents lists available at ScienceDirect

Journal of Equine Veterinary Science

journal homepage: www.j-evs.com

Original Research

Evaluation of Bacterial Contamination of Equine Cadaver Skin Following Injections Using Different Needle Types and Skin Preparation Methods

Tobias Puschmann [a,*], Jutta Verspohl [b], Karl Rohn [c], Bernhard Ohnesorge [a]

[a] *Clinic for Horses, Epidemiology and Information Processing, University of Veterinary Medicine Hannover, Foundation, Hannover, Germany*
[b] *Institute for Microbiology, Epidemiology and Information Processing, University of Veterinary Medicine Hannover, Foundation, Hannover, Germany*
[c] *Department of Biometry, Epidemiology and Information Processing, University of Veterinary Medicine Hannover, Foundation, Hannover, Germany*

ARTICLE INFO

Article history:
Received 8 February 2016
Received in revised form 28 June 2016
Accepted 28 June 2016
Available online 15 July 2016

Keywords:
Intramuscular injection complication
Skin core
Infection
Bacteria
Disinfection

ABSTRACT

To assess bacterial translocation through needle punctures and the impact of preparation protocols using equine cadaver skin and different needle sizes and types. Nonprepared, disinfected, clipped and disinfected, and clipped and antiseptically washed skin specimens were sampled sterilely using swabs and biopsies to compare disinfection efficacy. The skin specimens were punctured using 18G hollow needles (18G), 18G needles with stylet (18GM), and 22G hollow needles (22G) and were flushed with sterile saline solution. The saline was collected in blood culture systems or homogenization test tubes and processed for qualitative (ql) or semiquantitative microbiologic evaluation. Disinfection and antiseptic preparation of skin specimens led to a significant reduction of bacteria ($P < .05$) sampled with swabs and biopsies, whereas clipping did not produce further significant differences. Bacteria were detected in many needle punctures, depending on the preparation protocols and needle types and sizes used. The highest amount of bacteria was detected with 18GM and the lowest with 22G (ql; nonprepared; nonsignificant). Disinfection and antiseptic preparation reduced bacterial translocation significantly in 18G punctures (ql; $P < .05$), whereas clipping did not result in any further significant difference. Overall, more bacteria were detected after 18G needle punctures when samples were processed qualitatively. Disinfection before injections reduces the number of bacteria translocated into deeper tissue. Larger needles and needles with stylet increase the number of bacteria translocated through the skin.

1. Introduction

Infective complications following intramuscular (IM) injections include malignant edema and inflammatory processes, which cumulate in cellulitis and abscess formation [1]. Abscesses and cellulitis are frequent complications following IM injections with an incidence of 0.016% [2]. Furthermore, clostridial myonecrosis following IM injections is observed in horses [3–6]. The exact pathogenesis of clostridial infections following IM injections is unknown [5,7]. Injection abscesses are thought to develop either through an exogenic or endogenic pathway, with the latter being considered subordinate in human medicine [8]. Clostridial infections are thought to be initiated through clostridia being positioned on the skin, the needle, drug vials, or in the pharmaceutical itself [9]. On the other hand, it is thought that dormant spores, which are situated in equine muscle tissue and may be reactivated through traumatic influences, may cause clostridial myonecrosis [10].

* Corresponding author at: Tobias Puschmann, Clinic for Horses, University of Veterinary Medicine Hannover, Hannover 30559, Germany.
E-mail address: tobias.puschmann@tiho-hannover.de (T. Puschmann).

0737-0806/$ – see front matter © 2016 Elsevier Inc. All rights reserved.
http://dx.doi.org/10.1016/j.jevs.2016.06.079

When performing IM injections, pathogenic bacteria, which may adhere to skin cores or contaminations, can be advanced into deeper tissue layers through needle punctures [11]. The production of skin cores is considered unavoidable [12], and they are thought to be an initiative for infective complications following IM injections [12,13]. Skin cores are thought to be the primary carriers of pathogen bacteria because pharmaceuticals are produced sterilely. Therefore, complication rates may depend on the translocation of debris, skin, and hair fragments [11]. The introduction of bacteria with skin cores is not considered to necessarily cause clinically manifest injection site infections because physiologically immune defense mechanisms can eliminate infectious agents [14].

After antiseptic skin preparation, *Micrococci*, *Propionibacteria*, *Bacillus*, and *Clostridia* were isolated from needle contents after intra-articular knee punctures in human medicine [15,16]. In horses, positive bacterial cultures were obtained from 2.6% of needle contents after intra-articular punctures of antiseptically prepared joints [17]. Furthermore, an early study describes the fact that culturing needles also produce positive bacterial results after puncturing equine skin specimens. Setting the percentage of the bacterial results of unprepared skin at 100%, the study reports a reduction to 61% after clipping the skin specimens, to 54% after disinfection, and to 38% after clipping and disinfecting the skin specimens [18]. Contrary to the study of Koller [18], who describes a decrease of the bacterial population through clipping, Hague et al. [19] published results indicating that clipping equine skin specimens increases colony forming units. Nevertheless, there was no significant difference in colony forming units reported between clipped and nonclipped specimens after antiseptic preparation. Bacteria isolated from antiseptically prepared skin in this study included *Bacillus* sp., nonhemolytic *Staphylococcus* sp., and *Micrococcus* sp. The authors, therefore, conclude that clipping is not essential for antiseptic preparation [19].

Creating antiseptic conditions for IM injections is considered impracticable in equine medicine because complications occur very infrequently [20]. A cleaning or disinfection before IM injections, nevertheless, is strongly advised [14,18,21,22] to mechanically clean the injection site [23] and to reduce the amount of bacteria situated on the skin [24].

Therefore, one aim of this study was to investigate the influence of different skin preparation and disinfection methods on the number of bacterial isolates on equine skin specimens. Another aim was to investigate whether the number of bacterial isolates obtained from the needle's contents after skin puncture changes depending on the needle type and size and skin preparation method.

2. Materials and Methods

Skin specimens were obtained from horses which were euthanized at the Clinic for Horses and the Department of Physiology of the University of Veterinary Medicine, Hannover. All horses were euthanized for reasons independent of this study. Inclusion criteria demanded warmblood horses aged 1 year and older, presenting with average skin and hair condition. Horses included were had not been treated with antibiotics previously, were not suffering from skin diseases or hypovolemic/septic shock, and their hair coat was not secondarily contaminated or sweated.

Four differently prepared skin specimens were used for sample harvest in the hamstring region, approximately 15 cm distally of the ischial tuberosity: nonclipped and nondisinfected (no clip-no dis), nonclipped and disinfected once (no clip-dis), clipped and disinfected once (clip-dis), and clipped and antiseptically washed for 5 minutes (clip-as). Disinfection was performed using gauze with a 70% Ethanol solution (Softasept N, B. Braun Melsungen AG, Germany), and antiseptic washing was performed using gauze with chlorhexidine-gluconate (HiBiSCRUB, Regent Medical Overseas Limited, Great Britain). Chlorhexidine residues were cleaned with sterile saline solution (NaCl 0.9%, B. Braun Melsungen AG, Germany). All sampling procedures were performed antiseptically.

The skin surfaces of all four specimen types were sampled using a sterile swab system (eSwab, COPAN Italia S.p.A., Italy). The swab was wetted with its medium initially. The tip of the swab was positioned on each specimen and rotated 360°, using slight forward and backward movements. The hairs of the unclipped skin specimens were parted using a needle's tip to ensure that the tip of the swab made contact with both hair and skin surface. Swab samples were analyzed semiquantitatively using fractional smears to determine low, moderate, or high numbers of bacteria in the samples.

Skin biopsies from all four specimen types were obtained using a standard 6-mm biopsy punch (Biopsy Punch 6 mm, Stiefel Laboratorium GmbH, Germany). Skin biopsies were put into homogenization tubes (tubes: Firma Biozym, Scientific GmbH, Germany; glass balls: Fa Roth GmbH u. Co.KG, Germany) containing phosphate buffered saline (PBS) and were analyzed semiquantitatively.

The sampling areas were incised laterally on both sides and ventrally using a scalpel blade (Carbon Steel Skalpellklingen #22, B. Braun, Aesculap AG, Germany) for the production of skin cores. Both ventral corners were fixed using forceps to keep the specimen under traction while separating it from the underlying tissue and for the sampling procedure (Fig. 1A). Skin punctures were created with the subjectively extrapolated force and speed needed for performing IM injections. The needles were then flushed using 0.2 mL of sterile saline solution (Isotone Kochsalz-Lösung 0.9%, B. Braun Melsungen AG, Germany). The saline was collected in test tubes which were positioned underneath the needle's tip (Fig. 1B). Different samples were obtained, as depicted in Tables 1 and 2 (18G needles without Stylet (18G): BD Microlance3, Becton, Dickinson & Co. Ltd., Ireland, 18G needles with stylet (18GM): BD Spinal needle, Becton Dickinson S.A., Spain, 22G needles without stylet (22G): BD Microlance3, Becton Dickinson S.A., Spain). Different test tube systems were used (blood culture system: Oxoid SIGNAL Blood Culture System Medium, Oxoid Limited, England; homogenization tubes containing PBS or culture medium: tubes: Firma Biozym, Scientific GmbH, Germany; glass balls: Fa Roth GmbH u. Co.KG, Germany). The blood culture systems were stored at room temperature, whereas swab samples and homogenization tubes were stored at 4°C

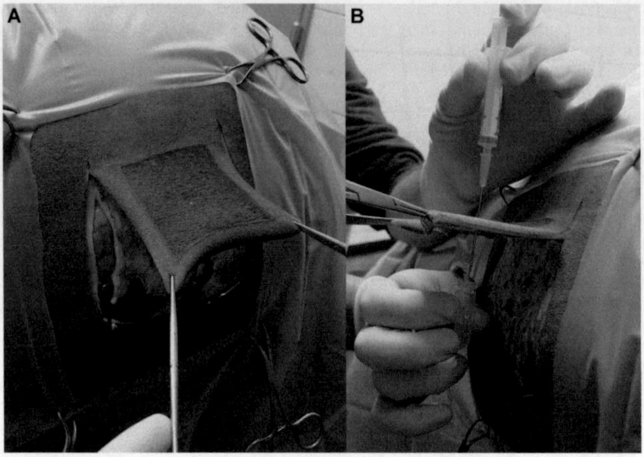

Fig. 1. Antiseptic skin preparation procedure and skin core harvest. (A) The skin specimen (no clip-no dis) is kept under tension using two forceps. (B) The skin specimen is punctured with an 18G needle and flushed with sterile saline. The saline is collected in a test tube.

until further processing. Samples were analyzed qualitatively or semiquantitatively (Tables 1 and 2).

Sample homogenization was performed at the Institute of Microbiology of the University of Veterinary Medicine, Hannover (Fast Prep FB 120, Model No.: FP120 HY-230, Fa. Hybaid, Germany), as well as microbiologic analysis, according to the institute's specific procedural and examination requirements. Fractional smears were used for semiquantitative evaluation, to differentiate low, moderate or high numbers of bacteria in the samples. Semiquantitative analysis results were converted into a scoring system for statistical evaluation (Table 3), whereas qualitative results were only interpreted as positive or negative for bacterial growth in general.

2.1. Statistical Analysis

Statistical analysis was performed using the statistic software SAS, Version 9.3 (SAS Institute, Cary, NC, USA). The Kolmogorov–Smirnov test and visual assessment of Q–Q plots was used for the assumption of normality. We applied distribution-free nonparametric methods because normal distribution was rejected for all data. One way analysis with repeated measurements were used for the comparison of the sampling devices within the four preparation protocols (skin swabs, skin biopsies and 18G needle's contents). Consequently, a permutation test with post hoc multiple pairwise comparisons and Sidak adjustment for *P*-values was used. Fisher's exact test was used to compare the 18G,

Table 1
Results of samples where semiquantitative evaluation was performed using different needles and processing methods, arranged by sample type and skin specimen type.

Sample Type	n	Preparation Method	Samples Containing Bacteria	Sum of Score	MEDIAN OF Score	Min of Score	Max of Score	Range of Score	Significance ($P < .05$)
Swab samples	10	no clip-no dis	10/10 (100%)	24	2	1	4	3	B, C, D
		no clip-dis	10/10 (100%)	10	1	1	1	0	A, D
		clip-dis	10/10 (100%)	14	1	1	3	2	A, D
		clip-as	2/10 (20%)	2	0	0	1	1	A, B, C
Skin biopsies	16	no clip-no dis	16/16 (100%)	37	2	1	4	3	B, C, D
		no clip-dis	15/16 (94%)	16	1	0	2	2	A, D
		clip-dis	15/16 (94%)	15	1	0	2	2	A, D
		clip-as	3/16 (19%)	3	0	0	1	1	A, B, C
18G needles	10	no clip-no dis	4/10 (40%)	4	0	0	1	1	Non
		no clip-dis	0/10 (0%)	0	0	0	0	0	Non
		clip-dis	1/10 (10%)	1	0	0	1	1	Non
		clip-as	0/10 (0%)	0	0	0	0	0	Non

Non, no statistical significance; preparation methods of 18G needles did not differ significantly from each other. Comparison of swab samples and skin biopsies were not significant in any combination.
Uppercase letters: significance within each sample type group: A, significant to no clip-no dis; B, significant to no clip-dis; C, significant to clip-dis; D, significant to clip-as.

Table 2
Results of samples where qualitative evaluation was performed using different needles and processing methods, arranged by sample type and skin specimen type.

Sample Type	n	Preparation Method	Samples Containing Bacteria	Significance ($P < .05$)
18G needle, hom	16	no clip-no dis	13/16 (81%)	B,C,D
		no clip-dis	6/16 (37%)	A
		clip-dis	3/16 (19%)	A
		clip-as	2/16 (12%)	A
18G needle blood culture	10	no clip-no dis	7/10 (44%)	non
18GM needle, hom	16	no clip-no dis	14/16 (87%)	non
22G needle, hom	16	no clip-no dis	8/16 (50%)	non

Abbreviations: A, significant to no clip-no dis; B, significant to no clip-dis; C, significant to clip-dis; D, significant to clip-as.
Non, no statistical significance; hom, homogenization tubes; needle sizes 18G (with blood culture or homogenization tubes), 18GM and 22G did not differ significantly from each other in an combination.

18GM and 22G needles' contents within no clip-no dis specimens and the processing procedure with either blood culture systems, culture medium, or PBS with 18G needles' contents. The two-sample signed rank test was used for the comparison of sampling devices within a specific skin specimen type (skin swabs and skin biopsies within each specimen type and PBS and culture medium samples of the 18G needle's contents). P values $< .05$ were considered statistically significant regarding all calculations.

3. Results

Sample acquisition was performed 15–30 minutes after euthanasia in all cases. The time in which samples were stored until further processing, ranged from 1–18.6 hours.

The incidence of bacterial detection and the total score number per sampling technique, as well as the results of mean, median, and range calculation are listed in Tables 1 and 2. The semiquantitative evaluation presented low numbers of bacteria in most cases and occasionally moderate numbers. High numbers were not detected in any sample.

The different bacterial species isolated from all samples are presented in Fig. 2, divided into Gram-positive and Gram-negative species and arranged according to their isolation frequency.

Similar result tendencies were observed when evaluating the impact of different skin preparation methods and disinfection protocols on the bacterial skin flora using swab samples (n = 10) and skin biopsies (n = 16), with the highest score numbers detected in samples taken from no clip-no dis specimens. Disinfection (no clip-dis) and clipping and disinfection (clip-dis) of skin specimens led to a significant reduction of score numbers in both sampling procedures ($P < .05$). There was no significant difference observed between no clip-dis and clip-dis specimens. Antiseptic skin preparation led to a further decrease of bacteria isolated with both sampling procedures, differing significantly to all other skin preparation methods ($P < .05$). No significant statistical difference was observed comparing the two sampling techniques (skin swabs, n1–n10; skin biopsies, n1–n10) within the individual skin preparation types (Table 1).

Evaluating the bacterial translocation through needle punctures, depending on the skin preparation and disinfection protocols, the highest numbers of isolates were also detected when no clip-no dis specimens were punctured with qualitative and semiquantitative evaluation, respectively. Compared to no clip specimens, the qualitative evaluation of 18G punctures revealed a significant reduction in score numbers if no clip-dis were punctured ($P <$ .05) and a further significant reduction with clip-dis and clip-as skin specimens ($P < .01$). By contrast, the semiquantitative evaluation of 18G punctures detected one isolate after puncturing clip-dis skin specimens and did not detect any bacteria in no clip-dis and clip-as specimens punctures (Table 1 and 2).

Comparing the different processing procedures using 18G needle punctures from no clip-no dis skin specimens, the highest number of bacterial isolates was detected if samples were processed for qualitative evaluation using the blood culture system. The second highest detection rate was observed with samples processed for qualitative evaluation using homogenization tubes filled with culture medium, and the lowest detection rate was found if samples were processed semiquantitatively using homogenization tubes filled with PBS. Nevertheless, the difference was not statistically significant.

Evaluation of the impact of needle size and type on the amount of bacterial isolates detected from no clip-no dis skin punctures showed that the highest number of isolates was detected using 18GM needles, followed by 18G needles. The lowest number of isolates was detected using 22G needles. The difference was not statistically significant.

Table 3
Scoring system used to evaluate semiquantitative bacterial results.

Score	Definition
0	no bacterial species detected
1	≤ 2 bacterial species, low numbers
2	≥ 2 bacterial species, low numbers
3	≤ 2 bacterial species, moderate numbers
4	≥ 2 bacterial species, moderate numbers

4. Discussion

We used skin specimens from freshly euthanized horses which presented clinically without skin diseases or other conditions, because these diseases may have had a modifying effect on the normal bacterial skin flora. We believe that using freshly harvested cadaver skin specimens did not

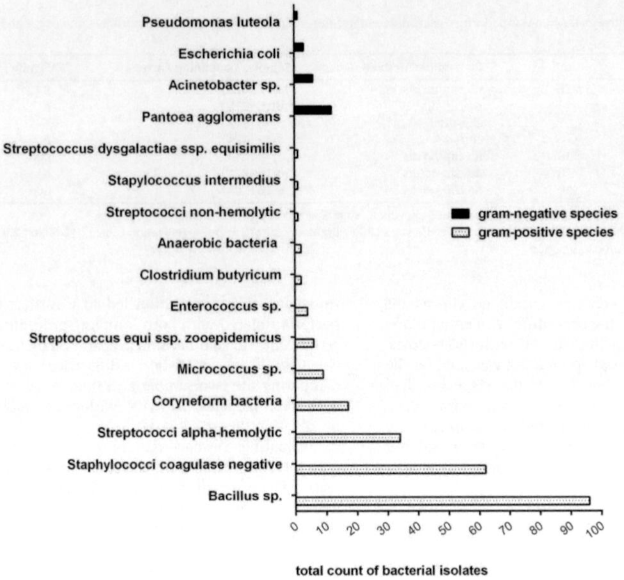

Fig. 2. Number of bacterial species isolated from all samples (swabs, skin biopsies and contents of different needle sizes and types). sp., species; ssp., subspecies.

have a significant impact on the bacterial skin flora we isolated because the period of time from euthanasia to sample acquisition was relatively short. The body location where skin specimens were harvested was chosen specifically as it is frequently used for IM injections [2]. The sampling procedure was performed antiseptically to avoid secondary sample contamination.

We slightly modified the method described previously by Koller [18], who used equine skin flaps of the neck region. We consider it a useful method, possibly without secondary contamination, because no bacteria could be isolated from most samples taken from antiseptically prepared skin specimens. The time period from sample acquisition until further microbiologic processing ranged from 1–18.6 hours, due to some samples being acquired at night time. Because samples were stored according to the standard guidelines of our hospital, we do not consider the time delay to be of major impact on the sample's composition and the isolation rate of bacteria.

The samples acquired from needle punctures were collected in blood culture systems and homogenization tubes containing culture medium. We hypothesized that homogenizing samples would result in detecting a larger amount of bacteria, especially those located deep within the tissue fragments. Skin biopsies were homogenized for the same reason. Semiquantitative evaluation was performed with certain samples to quantify the amount of bacteria being translocated by needle punctures better. We hypothesized that needle contents contain a very low number of bacteria, compared to 6-mm skin biopsies, and that culturing samples to induce growth would increase the chances of isolating bacteria from these samples. We planned to elucidate this hypothesis by comparing samples processed with PBS and culture medium.

We investigated skin preparation and disinfection methods using skin swabs to sample the superficial skin bacteria and skin biopsies to detect the effect of these procedures on the bacterial species which are located in deeper tissue layers. The results of swab samples and skin biopsies, respectively, presented similar tendencies. Disinfecting skin specimens once led to a significant decrease of bacterial isolates. Clipping the hair and disinfecting the skin specimens once did not produce significantly different results. The ethanol possibly disinfects and reaches the skin through the hair just as efficiently if the hair is not too long or visibly soiled. There was a further decrease observed in bacterial isolates when skin specimens were clipped and prepared antiseptically. The few bacteria species detected on antiseptically prepared skin specimens may represent bacteria from deeper tissue layers. On the other hand, antiseptic skin preparation is not considered to sterilize skin completely, it rather reduces the bacterial counts [25]. Therefore, isolating bacteria in low numbers from skin prepared antiseptically must be anticipated. We intended to detect bacteria located deeper within the skin, which would not be detected by only using skin swabs, by using skin biopsies. Nevertheless, there was no significant difference in the score number of isolates detected and bacteria species isolated from the sample groups. An explanation for these findings may be that only very small numbers of bacteria are located deep within the tissue, and most isolates had already been detected with swabs samples. Another possibility is that the homogenization process did not succeed in making these bacteria more accessible for our detection methods.

We used 18G needles on no clip-no dis skin specimens to compare the different processing procedures (blood culture system, homogenization tubes with PBS or culture medium) because the incidence of skin coring is reported to be higher with larger needle sizes [11,26]. More isolates were detected in the blood culture system group. We hypothesized initially that the homogenization procedure would increase the amount of bacteria detected. These findings may be due either to chance and may change with an increase of sample numbers or to the fact that the homogenizing of samples decreases the isolation frequency of bacteria with this method. We observed that puncture samples processed in homogenization tubes filled with PBS tend to have a lower isolation frequency, due possibly to containing a low number of bacteria and not performing enrichment procedures.

We used 18G needles to evaluate the influence of skin preparation and disinfection methods on the needle's contents. We used PBS samples to quantify the amount of bacteria being translocated via possibly existing skin cores and the needle itself. The same samples were also acquired and processed with culture medium, resulting in a higher isolation frequency of bacteria. The results of the PBS group only indicated that any disinfection and skin preparation method has a nonsignificant reducing effect on the number of bacteria translocated by needle punctures. The isolation numbers are, nevertheless, too low to draw any further conclusions. The samples of the culture medium group of 18G needle punctures provided results similar to those obtained from comparing disinfection methods using skin biopsies and swab samples. A reduction of the amount of bacteria being translocated via the needle punctures was observed. Therefore, we must assume that disinfection before injections reduces the risk of translocating bacteria significantly and may, therefore, reduce the risk of infective complications following IM injections. Nevertheless, it remains unclear with this method whether the translocated bacteria derive from actual skin cores or hair, or if they were carried on the needle's surface. The few bacteria isolated from puncturing antiseptically prepared skin specimens possibly resulted from bacteria which were located in deeper tissue layers and were therefore not made extinct by the disinfecting agents. Isolation of bacteria from the needle's contents after puncturing disinfected and/or antiseptically prepared skin specimens has been reported previously in equine and human orthopedic surgery [15–17].

We punctured no clip-no dis skin specimens to evaluate the effect of different needle sizes. The highest bacterial isolation frequency was to be expected with this skin specimen type, resulting in better comparability. We chose 18G and 22G hollow needles, because the range of these sizes is advised for use when IM injections are performed in equids [27,28]. Our results indicate that smaller needles translocate smaller numbers of bacteria. These results are to be expected, because the skin coring incidence is reported to be lower when smaller needle sizes are used [11,17,26,29], and the amount of bacteria being advanced through needle punctures is thought to depend on the amount of skin, hair, and debris being translocated [11]. Using an 18G spinal needle with stylet led to an increase of bacterial isolates. This may result from a higher skin coring incidence using these needles, as reported in one study [30], or an increase in hair translocation, reported in another study [26]. Hair fragments are reported to be translocated more easily by getting stuck between the stylet and the needle itself [17].

The bacteria isolated in this study are part of the skin flora or environmental or fecal contaminants. Bacterial species such as *Acinetobacter species*, *Staphylococcus intermedius*, nonhemolytic *Streptococci*, *Micrococcus species*, coryneform bacteria, coagulase negative *Staphylococci*, and *Micrococcus species* are documented to be part of the normal skin flora in equids [31]. Anaerobic bacteria are found in large numbers in follicles in human skin and are considered part of the skin flora [32]. *Staphylococcus intermedius* may be associated with subcutaneous abscess formation and wound infections in horses [31]. Some of the species, such as *Escherichia coli*, *Enterococci* (Pantoea agglomerans formerly named Enterobacter agglomerans), and *Streptoccoci* subspecies, cause cellulitis in horses infrequently [31]. *Streptococcus equi subspecies zooepidemucus* is a commensal found on mucus membranes, whereas *Streptococcus dysgalactiae subspecies equisimilis* is found on mucus membranes and equine skin; both may cause opportunistic infections [33]. *Pseudomonas luteola* is described as causing infections in humans [34], but its significance in equids is unclear. Alpha-hemolytic *Streptococci* were also found in equine placentitis [35] and the tracheas of horses with pulmonary disease [36], but their clinical significance remains unclear. *Clostridium butyricum* is generally considered nonpathogenic, but its spores were previously isolated from equine skeletal musculature [10]. Their role, especially considering the etiology of clostridial myonecrosis, is unclear, but dormant clostridia are reported to cause infections in cattle after being reactivated through traumatic influences [37]. We have isolated these clostridia on equine skin. Therefore, the study published by Vengust et al. [10] and ours give rise to the question of whether the internal or external pathway, as discussed previously in literature [8], is the common route of infections following injection. We consider it possible that other clostridia species or any pathogenic bacteria may be situated on equine skin as contaminants and may, therefore, be translocated into deeper tissue layers through injections. They may not necessarily cause an infection because their numbers are usually too low, as proven in this study. Nevertheless, their role as an infective agent must be considered in combination with compromised tissue integrity and a reduction of local immune defense mechanisms caused by the tissue toxicity of the drugs administered.

We conclude that performing a single disinfection before IM injections is advisable, because it reduces the number of bacteria being translocated via needle punctures. Therefore, the risk of causing an infective injection complication will be lowered. This is supported by a survey, which indicates that adverse drug reactions occur less frequently if disinfection is performed before IM injections [2]. Clipping the hair in addition to a single disinfection in patients presenting with average hair coat and skin condition is not necessary. Performing antiseptic site preparation is time-consuming. Therefore and because of the low frequency of complications following IM injections, it seems unpractical. Nevertheless, it can be considered in patients with visibly soiled hair and skin as it leads to a further reduction of the bacterial load.

In conclusion, disinfection before injections reduces the number of bacteria translocated into deeper tissue. Larger needles and needles with a stylet increase the number of bacteria translocated through the skin. Therefore, the translocation of bacteria can be reduced by using smaller needle sizes and needles without stylets.

Acknowledgments

The authors acknowledge the contributions of all coworkers at the Institute for Microbiology, University of Veterinary Medicine, Hannover, for sample processing, Dr. Klaus Hopster DipECVAA for writing assistance and all coworkers at the Clinic for Horses, University of Veterinary Medicine, Hannover, for assistance with sample acquisition.

All authors disclose that there are no conflicts of interest which inappropriately influence/bias this work.

References

[1] Koch G. Zur Haftung des Tierarztes bei Injektionen unter besonderer Berücksichtigung der Desinfektion [Dissertation]. Berlin: Humboldt-Universität zu Berlin; 1966.
[2] Ohnesorge B, Pfalzgraf S, Rohn K, Neuhaus M, Deegen E. Unverträglichkeitsreaktionen nach intramuskulärer Injektion beim Pferd - Auswertung einer Tierärztebefragung. Pferdeheilkunde 2006;22: 337–46.
[3] Peek SF, Semrad SD, Perkins GA. Clostridial myonecrosis in horses (37 cases 1985-2000). Equine Vet J 2003;35:86–92.
[4] Valberg S, McKinnon A. Clostridial cellulitis in the horse: a report of five cases. Can Vet J 1984;25:67–71.
[5] Rebhun W, Shin S, King J, Baum K, Patten V. Malignant edema in horses. J Am Vet Med Assoc 1985;187:732–6.
[6] Breuhaus B, Brown C, Scott E, Ainsworth D, Taylor R. Clostridial muscle infections following intramuscular injections in the horse. J Equine Vet Sci 1983;3:42–6.
[7] Peek SF, Semrad SD. Clostridial myonecrosis in horses. Equine Vet Educ 2002;14:163–8.
[8] Gabka J. Komplikationen der intramuskulären Injektion und ihre Behandlung. In: Gabka J, editor. Die Injektion; Technik, Praxis, Komplikationen. 1st ed. Berlin: Walter de Gruyter & Co.; 1968. p. 117–29.
[9] Brown CM, Kaneene JB, Walker RD. Intramuscular injection techniques and the development of clostridial myositis or cellulitis in horses. J Am Vet Med Assoc 1988;193:668–70.
[10] Vengust M, Arroyo R, Weese JS, Staempfli H, Baird J. Clostridial myositis: evaluation of normal equine skeletal muscle for the presence of clostridial spores. In: 48th Annual Conference American association of equine Practitioners. Orlando, Florida: Reprinted in IVIS with the permission of the AAEP; 2002. p. 134–5.
[11] Willener AW. Mechanik der Injektion und Verunreinigung des Stichkanals [Dissertation]. Bern: University of Bern; 1948.
[12] Haindl H, Klinge O. Post injection abscess caused by non-suitable cannulas? Biomed Eng 1989;34:268–71.
[13] Gibson T, Norris W. Skin fragments removed by injection needles. Lancet 1958;2:983–5.
[14] Ammann K. Die forensische Bedeutung der Injektionsschäden. Schweiz Arch Tierheilkd 1954;96:569–79.
[15] Bernau A, Cornelius C, Dauber W, Dietrich G, Heeg P. Hautstanzzylinder bei Gelenkpunktionen. Orthop Praxis 1985;5:359–65.
[16] Heeg P, Bernau A, Dauber W. Infektionsgefahr durch Injektionen. Biomed Eng 1990;35:179–81.
[17] Wahl K, Adams SB, Moore GE. Contamination of joints with tissue debris and hair after arthrocentesis: the effect of needle insertion angle, spinal needle gauge, and insertion of spinal needles with and without a stylet. Vet Surg 2012;41:391–8.
[18] Koller A. Injektion und Infektion [Dissertation]. Zürich: Veterinärmedizinische Fakultät, Universität Zürich; 1947.
[19] Hague BA, Honnas CM, Simpson RB, Peloso JG. Evaluation of skin bacterial flora before and after aseptic preparation of clipped and nonclipped arthrocentesis sites in horses. Vet Surg 1997;26:121–5.
[20] Williams PD. Intramuscular injection technique. Equine Vet Educ 1995;7:86–8.
[21] Zeller R. Über die Haftpflicht des Tierarztes bei Injektionen. Prakt Tierarzt 1958;39:10–3.
[22] Deegen E. Intramuskuläre Injektionen und Komplikationen. 10 Bpt-Kongress. Münster: Vortragsband Kleintiere, Pferd; 2003. p. 100–1.
[23] Gerber H, Tercier P, Müller A. Injektionstechnik - Injektionsfolgen beim Pferd. Schweiz Arch Tierheilkd 1980;122:205–15.
[24] Eikmeier H. Injektionen. In: Eikmeier H, Fellmer E, Moegle H, editors. Lehrbuch der gerichtlichen Tierheilkunde. 1st ed. Berlin und Hamburg: Verlag Paul Parey; 1990. p. 88–95.
[25] Aly R, Malbach HI. Comparative antibacterial efficacy of a 2-minute surgical scrub with chlorhexidine gluconate, povidone-iodine, and chloroxylenol sponge-brushes. Am J Infect Control 1988;16:173–7.
[26] Puschmann T, Haeger JD, Hambruch N, Rohn K, Pfarrer C, Ohnesorge B. Histological evaluation of equine cadaver skin cores dependent on needle type and skin preparation method to investigate intramuscular injection complications in horses. J Equine Vet Sci 2016;45:8–16.
[27] Rose R, Hodgson D. Medication administration. In: Rose R, Hodgson D, editors. Manual of equine practice. 2nd ed. Philadelphia: W.B. Saunders Company; 2000. p. 752–6.
[28] Rossdale P, Ricketts S. Equine stud farm medicine. 2nd ed. London: Bailliere Tindall/Cassell Ltd.; 1980.
[29] Adams SB, Moore GE, Elrashidy M, Mohamed A, Snyder PW. Effect of needle size and type, reuse of needles, insertion speed, and Removal of Hair on Contamination of Joints with Tissue Debris and Hair after arthrocentesis. Vet Surg 2010;39:667–73.
[30] Charlebois PA. Coring: the unseen menace. Can Anaesth Soc J 1966; 13:585–97.
[31] Scott DW, Miller WHJ. Equine dermatology. 2nd ed. Maryland Heights, Missouri: Elsevier Health Sciences; 2010.
[32] Puhvel SM, Reisner RM, Amirian DA. Quantification of bacteria in isolated pilosebaceous follicles in normal skin. J Invest Dermatol 1975;65:525–31.
[33] Timoney JF. The pathogenic equine streptococci. Vet Res 2004;35: 397–409.
[34] Chihab W, Alaoui AS, Amar M. Chryseomonas luteola identified as the source of serious infections in a Moroccan University Hospital. J Clin Microbiol 2004;42:1837–9.
[35] Hong C, Donahue J, Giles R, Petrites-Murphy M, Poonacha K, Roberts A, et al. Etiology and pathology of equine placentitis. J Vet Diagn Invest 1993;5:56–63.
[36] Benson CE, Sweeney C. Isolation of Streptococcus pneumoniae type 3 from equine species. J Clin Microbiol 1984;20:1028–30.
[37] Richey EJ. Clostridial (blackleg) diseases of cattle. Gainesville, Florida: University of Florida Cooperative Extension Service, Institute of Food and Agriculture Sciences, EDIS; 1996. p. 1–3.

3 Übergreifende Diskussion

In der vorliegenden Studie wurde der Einfluss von Kanülenart und Größe sowie Vorbereitungsart und Desinfektion der Injektionsstelle auf die Inzidenz (d.h. die Häufigkeit der Entstehung von Hautstanzen bei Punktionen) und die Anzahl und Größe der Hautstanzen je Punktion (regelmäßig wurden mehrere Haut- und Haarfragmente nach einer einmaligen Punktion beobachtet) untersucht. Weiterhin wurden die histologische Beschaffenheit der Hautstanzen und deren bakterieller Keimgehalt evaluiert.

3.1 Die Wirksamkeit von vorbereitenden Maßnahmen (Reinigung, Desinfektion) an der Injektionsstelle vor einer Injektion

3.1.1 Desinfizierende Reinigung

Eine Desinfektion und/oder Reinigung der Injektionsstelle vor einer i.m. Injektion wird dem heutigen medizinischen Standard entsprechend als obligatorisch erachtet (KOLLER 1947; AMMANN 1954; ZELLER 1958; DEEGEN 2003; GPM Leitfaden 2013). In der vorliegenden Studie konnte eine signifikante Reduktion der Bakterienpopulation durch die einfache, desinfizierende Reinigung der Haut nachgewiesen werden. Der Nachweis gelang durch eine Beprobung der Hautpräparationen mit Hauttupfer und Hautbiopsien. Bisher wurde in Studien zwar der Effekt einer antiseptischen, chirurgischen Vorbereitung von Punktionsstellen auf eine mögliche Keimbelastung untersucht (HAGUE et al. 1997; ZUBROD et al. 2004). Vergleichende Studien unter Berücksichtigung der in der Praxis üblichen, desinfizierenden Reinigung der Haut auf eine mögliche Keimverschleppung ins Gewebe wurden jedoch bisher nicht durchgeführt. In einer älteren Studie wurde lediglich der Einfluss von Desinfektion bzw. Scheren der Haut auf die Oberflächenkontamination von seinerzeit gebräuchlichen Mehrweg-Stahlkanülen untersucht (KOLLER 1947).

Eine weitere signifikante Reduktion der Keimpopulation wurde in dieser Studie durch die antiseptische Waschung erzielt. Diese Vorbereitung führt zwar zu weiterer Keimminderung, setzt allerdings eine Kürzung des Haarkleides voraus (s.u.). Das Risiko, Keime durch den Injektionsvorgang in tiefere Gewebeschichten zu verschleppen, erwies sich somit nach der antiseptischen Vorbereitung als am geringsten. Wegen des Zeitaufwandes und der geringen Komplikationsfrequenz ohne antiseptische Vorbereitung, erscheint diese Vorbereitungsmethode jedoch als nicht praktikabel oder erforderlich. Dennoch sollte eine, die einfache Desinfektion übersteigende, Reinigung und Desinfektion der Injektionsstelle bei verschmutztem und langem Haarkleid in Betracht gezogen werden.

3.1.2 Kürzen des Haarkleides

Zur Unterstützung einer desinfizierenden Reinigung der Injektionsstelle wird die Kürzung der Haare bei langem Haarkleid angeraten (DEEGEN 2003; GPM Leitfaden 2013). Zur Effizienz einer derartigen Kürzung konnte für die antiseptische Vorbereitung der Haut allerdings kein signifikanter Unterschied zwischen geschorenem und nicht geschorenem Haarkleid beobachtet werden (HAGUE et al. 1997). In der vorliegenden Studie konnte auch für die einfache, desinfizierende Reinigung kein signifikanter Unterschied im Keimgehalt bei nicht geschorenen Haaren im Vergleich zu geschorenen Haaren nachgewiesen werden. Berücksichtigt werden muss jedoch, dass alle Probanden einen durchschnittlichen guten Pflegezustand aufwiesen, und ein Vergleich zu übermäßig langem oder verschmutztem Haarkleid in dieser Studie aufgrund des Probenmaterials nicht durchgeführt werden konnte. Der Nachweis, dass bei langem, verschmutztem Haarkleid die Kürzung der Haare die Keimbelastung reduziert war somit nicht zu führen. Da bei langem Haarkleid eine schlechtere Wirkung der Desinfektionsmittel zu erwarten ist, bleibt die Kürzung langer Haare bis auf weiteres dennoch empfehlenswert.

3.2 Der Einfluss einer Kürzung der Haare auf die Entstehung von Hautstanzen und Haarpartikeln

3.2.1 Hautstanzen

Neben den Effekten des Scherens auf die Effizienz einer Desinfektion, wurde in der vorliegenden Studie ebenfalls der Einfluss dieser Maßnahme auf das Entstehen von Hautstanzen und die Verschleppung von Haarpartikeln berichtet. Bereits WILLENER (1948) und WAGENKNECHT (1924) beobachteten eine vermehrte Hautstanzinzidenz bei geschorener oder rasierter Haut und sprachen sich gegen diese Art der Injektionsstellenvorbereitung aus. Im Gegensatz hierzu publizierten WAHL et al. 2012 Ergebnisse, die auf eine Reduktion der Hautstanzinzidenz deuten, wenn geschorene Haut punktiert wird. In der vorliegenden Studie war dazu lediglich nachweisbar, dass die Inzidenz von Hautstanzen nur bei Verwendung der 18G Kanülen mit Stilett signifikant anstieg. Die Zunahme bei der Punktion mit 22G Kanülen ohne Stilett war nur tendenziell. Bei allen verwendeten Kanülenarten erwies sich der Unterschied von Anzahl und Größe der Hautstanzen je Punktion zwischen geschorener und nichtgeschorener Haut als nicht signifikant.

Betrachtet man zur Inzidenz und Anzahl und Größe der Hautstanzen je Punktion auch deren Keimbelastung, so war zwischen nicht geschorener und desinfizierter Haut und geschorener und desinfizierter Haut kein signifikanter Unterschied festzustellen. Demnach war in der vorliegenden Studie kein negativer Effekt des Scherens auf den Keimeintrag, die Hautstanzinzidenz, und die Größe und Anzahl der Hautstanzen je Punktion nach einmaliger Desinfektion mit Kanülen (ohne Stilett) nachweisbar. Ein nachteiliger Effekt der Schur bei der Injektionsvorbereitung, wie er aus der geschätzt erhöhten Häufigkeit von UAW nach Schur zu erwarten wäre (OHNESORGE et a. 2006), lässt sich demnach aus dieser Studie nicht ableiten.

3.2.2 Haarpartikel

Auch die Verschleppung von Haarpartikeln durch Punktionen wurde in dieser Studie untersucht. Ähnlich wie in der Studie von WILLENER (1948), führte das Scheren der Haut bei allen hier verwendeten Kanülenarten zu einer Erhöhung der Inzidenz von verschleppten Haarpartikeln, die allerdings nicht signifikant war. Eine derartige Zunahme könnte durch kleine, freie Haarfragmente bedingt sein, die bei der Schur entstehen. Außerdem wäre es möglich, dass kürzere Haarstummel eher in das Kanülenlumen hineinragen, sodass Haarstummel leichter mit Hautstanzen gemeinsam verschleppt werden. Auch hier führte bei geschorener und ungeschorener Haut die Verwendung größerer Kanülen tendenziell zu einer erhöhten Anzahl von verschleppten Haarpartikeln.

So war die größte Kontamination mit Haarpartikeln nach Punktion mit 18G Kanülen mit Stilett bei geschorener Haut festzustellen. Dies ist vergleichbar mit dem Ergebnis einer experimentellen Studie zu Gelenkspunktionen, in der Haare zwischen Stilett und Nadel eingeklemmt und so durch die Haut verschleppt wurden (WAHL et al. 2012). Eventuell verkeilen sich die kurz geschorenen Haare leichter zwischen Stilett und Nadel. Ob die erhöhte Anzahl von Haarpartikeln bei Verwendung von 18G Kanülen mit Stilett im Vergleich zur gleichen Kanülengröße ohne Stilett auch zu einem höheren Keimeintrag bei geschorenen Hautpräparationen führt, wurde in dieser Studie nicht untersucht. Die Bakteriennachweise bei Punktionen mit 18G Kanülen ohne Stilett bei geschorener Haut waren allerdings nicht höher als nach Punktionen der ungeschorenen Haut. Die dabei zu beobachtende Zunahme von verschleppten Haarfragmenten war nicht signifikant. Dabei ist jedoch zu berücksichtigen, dass Haarfragmente allein das Auftreten aseptischer Fremdkörperreaktionen hervorrufen können. Solche Reaktionen wurden vereinzelt nach intravenösen Punktionen bei Menschen, Hunden, Kaninchen, Ratten und Mäusen beschrieben. Die hierbei entstandenen Hautstanzen gelangten über den Blutstrom in die Lunge und lösten dort eine lokale Fremdkörperreaktion aus (INNES et al. 1958; TEKELI 1974; ANDREW 1976; SCHNEIDER u. PAPPRITZ 1976; KAST u. TSUNENARI 1983; MORTON et al. 1997). Eine klare Trennung zwischen Haarfragmenten und Hautstanzen war histologisch nicht immer möglich, da bei einigen Hautstanzen die Haarfollikel und der Haarschaft eines Haarfragments im Gewebe integriert waren. Die Frage, welche der beiden Strukturen einen höheren Keimeintrag verursachen, konnte daher nicht sicher beantwortet werden. WILLENER (1948) diskutiert, dass Haarstummel durch ihre glatte Oberfläche weniger Bakterienanhaftung erlauben als Hautstanzen. Außerdem enthielten Hautstanzen mehr bakterienbesiedelte Strukturen wie z.B. Talgdrüsen. Der Autor geht daher davon aus, dass die Verschleppung von Haarfragmenten eine untergeordnete Bedeutung für den Keimeintrag hat.

3.3 Die histologische Zusammensetzung von Hautstanzen

In der vorliegenden Studie konnten in den Paraffinschnitten der Hautstanzen sämtliche Strukturen der Haut erkannt werden, die mit Bakterien besiedelt sein können. So waren in einzelnen Schnitten Bestandteile des *Stratum corneum* der Epidermis, vitale Schichten der Epidermis, isolierte Haarschäfte, Haarschäfte, die in die Epidermis integriert waren, Haarwurzeln, Schweiß- und Talgdrüsen und Bindegewebe vorhanden. Hautstanzen müssen daher grundsätzlich als eine mögliche, im Einzelfall unvermeidliche Infektionsquelle angesehen werden.

3.3.1 Der Einfluss der Kanülengröße auf die Verschleppung von Hautstanzen

WILLENER (1948) diskutiert, dass die Menge des bakteriellen Eintrags von der Menge an verschleppten Hautpartikeln abhängt. So stellte er in seinen Untersuchungen eine zunehmende Verunreinigung mit steigendem Kanülendurchmesser fest. Die Ergebnisse unserer Studie unterstützen seine Hypothese. Auch hier war die Hautstanzinzidenz, sowohl bei geschorener als auch nicht geschorener Haut, bei der Benutzung kleinerer Kanülendurchmesser verringert. Zusätzlich nahm die Anzahl und Größe der Hautstanzen je Punktion signifikant ab, wenn kleinere Kanülen benutzt wurden. Korrelierend hierzu wurden weniger Bakterien in den Kanüleninhalten nachgewiesen, wenn kleinere Kanülen verwendet wurden. Diese Ergebnisse legen nahe, dass eine geringere Anzahl und Größe der Hautstanzen je Punktion zu einer geringeren Keimbelastung führen. Die Benutzung kleinstmöglicher Kanülen für die i.m. Injektion stellt daher eine Maßnahme dar, die möglicherweise die Inzidenz von Komplikationen nach Injektionen senken kann (z.B. abszedierende oder phlegmonöse Entzündungen). Dies wäre ebenfalls auf das Risiko von aseptischen Entzündungsreaktionen zu übertragen, die als Fremdkörperreaktion durch Hautstanzen verursacht werden können.

3.3.2 Der Einfluss der Anwendung von Kanülen mit Stilett auf die Entstehung von Hautstanzen

In der vorliegenden Studie war die Inzidenz von Hautstanzen bei Verwendung gleicher Kanülengröße (18G) mit Stilett an ungeschorener Haut signifikant geringer als ohne Stilett. Die Anzahl und Größer der Hautstanzen je Punktion von 18G Kanülen mit Stilett fiel signifikant niedriger aus als bei 18G Kanülen ohne Stilett. Allerdings führte das vorbereitende Scheren zu einer signifikanten Zunahme der Hautstanzinzidenz, wenn Kanülen mit Stilett verwendet wurden (s.o.). Hingegen führte das Scheren, bezogen auf die Größe und Anzahl der Hautstanzen je Punktion, bei beiden Kanülenarten nicht zu einem signifikanten Unterschied. Die Keimkontamination war bei den 18G Kanülen mit Stilett höher (nicht signifikant) als bei Verwendung von Kanülen ohne Stilett. Andere Studien, welche die Keimkontamination durch Kanülen mit Stilett und Kanülen ohne Stilett untersucht haben, liegen nicht vor. Erklären ließe sich dieser zunächst überraschende Effekt durch das Ausreißen und Einklemmen der Haare zwischen Stilett und Kanüle, was zur Eröffnung von mehr Haarbälgen und anhaftenden Talgdrüsen führen könnte. Ebenso könnte sich so bei geschorener Haut die Hautstanzinzidenz erhöhen, da viele Haarfragmente in Gewebefragmente integriert sind. Trotz der signifikant erhöhten Inzidenz, veränderte sich aufgrund des Stiletts die Anzahl und Größe der Hautstanzen je Punktion nach dem Scheren nicht signifikant. Ob die Zunahme der Hautstanzinzidenz bei 18G Kanülen mit Stilett bei geschorener Haut einen Einfluss auf Injektionskomplikationen hat, bleibt allerdings fraglich.

3.4 Die Bedeutung der semiquantitativ oder qualitativ nachgewiesenen Bakterien auf die Entstehung einer Injektionskomplikation

Die semiquantitative bakteriologische Auswertung der Hautstanzen in unserer Studie ergab ausnahmslos geringgradige Keimnachweise. Zusätzlich wurden in den semiquantitativ analysierten Proben (ohne Keimanreicherung) insgesamt weniger Bakterien isoliert als in den qualitativ analysierten Proben (mit Keimanreicherung). Daher ist anzunehmen, dass Hautstanzen von 18G Kanülen eine sehr geringe Keimzahl enthalten. Ob die Menge an Bakterien ausreichend ist um eine Infektion zu initiieren ist fraglich. Es ist davon auszugehen, dass dafür mehrere Faktoren verantwortlich sind. In gut durchbluteten Muskelbereichen muss angenommen werden, dass die natürliche Immunreaktion kompetent genug ist, eine Infektion abzuwenden.

Ist jedoch bei einer Injektion die lokale Immunkompetenz herabgesetzt, könnten Hautstanzen eine Infektion verursachen. Vorstellbar ist, dass die lokale oder systemische Immunkompetenz durch Gewebenekrosen aufgrund gewebetoxischer Medikamentenwirkungen durch Hämatombildung oder eine Grunderkrankung vermindert wird (HAINDL u. KLINGE 1989).

In einer solchen Abwehrlage ist anzunehmen, dass das geschädigte Gewebe wie eine Anreicherungskultur fungiert, in der Keime auf günstige Umgebungsbedingungen stoßen, die eine Vermehrung und letztendlich die Etablierung einer klinisch manifesten Infektion begünstigen.

3.5 Übertragbarkeit der Ergebnisse auf andere Injektionsarten

Die Ergebnisse der vorliegenden Studie zum Vergleich der Effizienz von Desinfektions- und Vorbereitungsmaßnahmen der Injektionsstelle sind auf andere Injektionsarten, wie subkutane (s.c.) Injektionen und intravenöse (i.v.) Injektionen vermutlich teilweise übertragbar. Hautstanzen, die bei s.c. Injektionen entstehen und in das Gewebe verbracht werden, können gegebenenfalls leichter eine Infektion verursachen, da anzunehmen ist, dass die subkutanen Gewebe weniger gut durchblutet sind. Auf der anderen Seite werden für subkutane Injektionen kleinlumige Kanülen verwendet, wodurch auch nach dieser Studie die Keimbelastung, Inzidenz, Größe und Anzahl an Hautfragmenten je Injektion deutlich vermindert wird. Dies wiederum würde die in der Literatur vermutete, niedrigere Inzidenz an bakteriellen Injektionskomplikationen erklären. Bei i.v. Injektionen werden möglicherweise vorhandene Hautstanzen durch das Durchspülen der Kanüle mit Blut nach Stauung entfernt, bevor die Injektion getätigt wird. Wird direkt mit aufgesetzter Spritze aspiriert und injiziert, gelangen potentiell vorhandene Hautstanzen nicht in das Gewebe, sondern in den Blutkreislauf. So wurden in anderen Studien/Berichten bei Mäusen, Ratten, Kaninchen, Hunden und Menschen Hautfragmente in die Lunge abgespült, wo sie Fremdkörperreaktionen auslösen können (INNES et al. 1958; TEKELI 1974; ANDREW 1976; SCHNEIDER u. PAPPRITZ 1976; KAST u. TSUNENARI 1983; MORTON et al. 1997). Ob hierdurch auch eine bakterielle Komplikation oder nachhaltige Schädigung des Patienten erfolgt, bleibt bisher unbekannt. WILLENER (1948) beschreibt in seinen Versuchen, dass einige Hautstanzen nicht vollständig ausgestanzt werden und lediglich nach innen in den Stichkanal umgeklappt werden. Auch waren Haarfragmente im Stichkanal nachzuweisen. Dies könnte bei i.v. Injektionen möglicherwiese eine bakterielle Infektion des Stichkanals ermöglichen und zur Entstehung einer Phlebitis, Peri- oder Thrombophlebitis beitragen. Kanülen mit Stilett werden in der Regel nicht für i.m., s.c. oder i.v. Injektion verwendet und dienten in dieser Studie lediglich zur Überprüfung des Unterschiedes zu Hohlnadeln. Diese Kanülenart wird in der Regel bei Gelenkspunktionen oder

Liquorpunktionen eingesetzt. In diesen Fällen wird der Injektionsbereich geschoren oder rasiert und antiseptisch vorbereitet. Ob bei der antiseptischen Vorbereitungsart ebenfalls eine höhere Inzidenz und Keimbelastung bei Verwendung von Kanülen mit Stilett zu erwarten ist, wurde in dieser Studie nicht untersucht. In einer anderen Studie wurde lediglich eine bakteriologische Untersuchung von 16 G Kanüleninhalten mit Hautstanzen (Hohlnadeln) nach der Punktion von antiseptisch vorbereiteter Haut durchgeführt. Hier wurden in 2,61% der Proben Bakterien nachgewiesen (WAHL et al. 2012). Die Autoren diskutieren, dass die Bakterien von Haarfollikeln und Talgdrüsen in der Haut stammen, da auf zuvor entnommenen Hauttupfern kein Keimwachstum nachgewiesen wurde. Eine Möglichkeit, die Verschleppung von Hautfragmenten mit diesen Kanülen zu reduzieren, wäre eine vor der Punktion durchgeführte, kleine Inzision der Haut, durch welche die Nadel dann eingeführt werden kann.

3.6 Zusammenfassung

Zusammenfassend ist festzustellen, dass Hautstanzen histologisch sämtliche Strukturen der Haut enthalten, die von Bakterien besiedelt sind. Daher ist eine Beteiligung an der Entstehung bakterieller Infektionen (oder Fremdkörperreaktionen) nach Injektionen denkbar. Schon die einmalige Desinfektion der Haut kann eine signifikante Reduktion der Bakterienpopulation erzielen. Auf diese Weise kann das Risiko einer Keimverschleppung reduziert werden. Da die Verwendung kleiner Kanülen zu einem geringeren Keimeintrag und zu einer geringeren Hautstanzinzidenz, Größe und Menge an Hautstanzen je Punktion führt, sollten in Abhängigkeit von der Konsistenz des Injektats die möglichst kleinlumigsten Kanülen verwendet werden.

4 Zusammenfassung

Tobias Puschmann:

Histologische und mikrobiologische Untersuchung bei Punktionen entstehender Hautstanzen, mit verschiedenen Kanülenarten und Desinfektions- und Vorbereitungsmethoden der Injektionsstelle beim Pferd

Das Ziel dieser Studie war die Untersuchung des Einflusses verschiedener Desinfektions- und Vorbereitungsmethoden der Haut, auf die Bakterienflora der Injektionsstelle und die Bakterienflora der Hautausstanzungen, die durch Punktionen entstehen können. Zusätzlich wurde das Auftreten von Hautstanzen untersucht und deren histologische Bestandteile evaluiert.

Verschiedene Hautpräparationen, nicht geschoren und nicht desinfiziert (*no-clip*), nicht geschoren und desinfiziert (*dis.no-clip*), geschoren und desinfiziert (*dis.clip*) und geschoren und antiseptisch gewaschen (*as.clip*), wurden von frisch euthanasierten Pferden gewonnen. Für die Produktion von Hautstanzen, wurden die Hautpräparationen mit 18G Kanülen (18G), 18G Kanülen mit Stilett (18GM) und 22G Kanülen (22G) punktiert. Die Kanülen wurden mit steriler Kochsalzlösung gespült.
Die Wirksamkeit der Desinfektions- und Vorbereitungsprotokolle wurde mittels Hauttupfer und Hautbiopsien untersucht. Für die mikrobiologische Untersuchung der Hautstanzen wurde die Spüllösung in Blutkultursysteme oder Probenröhrchen, die entweder phosphatgepufferte Kochsalzlösung oder Nährmedium enthielten, installiert. Es wurden qualitative und semiqualitative Bakteriennachweise durchgeführt.
Für die mikroskopische Auswertung der Inzidenz der Hautstanzen und der Größe und Anzahl von Hautstanzen je Punktion, wurde die Spüllösung auf Objektträgern getrocknet und eine Färbung nach Pappenheim durchgeführt. Zusätzlich wurden Hautstanzen, die mit 18G Kanülen angefertigt wurden, in 4% Formalin fixiert und anschließend in Paraffin eingebettet und geschnitten. Die Paraffinschnitte wurden einer Massonfärbung unterzogen. Die histologische Auswertung erfolgte mittels Lichtmikroskopie.

Allumfassend wurden verschiedene Bakterienarten, wie Hautkommensalen und Bakterien der Darmflora nachgewiesen. Im Vergleich der Desinfektionsprotokolle wurden signifikant weniger Bakterien nachgewiesen, wenn die Hautpräparationen desinfiziert oder antiseptisch vorbereitet wurden ($p < 0{,}05$), wohingegen Desinfizieren und Scheren keine zusätzliche Reduktion der Bakteriennachweise erzielte. Vergleichbare Ergebnistendenzen der Bakteriennachweise wurden beobachtet, wenn 18G Kanüleninhalte nach der Punktion der verschiedenen Hautpräparationen untersucht wurden ($p < 0{,}05$). Kanülenartvergleichend wurden die niedrigsten Bakteriennachweise nach Punktionen mit 22G Kanülen erzielt und die höchsten Nachweise nach der Punktion mit 18GM Kanülen (qualitativ; *no-clip*; nicht signifikant).
Abhängig von den Kanülenarten und den Präparations- und Desinfektionsmethoden wurden nach vielen Punktionen Hautstanzen nachgewiesen. Hautstanzen wurden übergreifend häufiger beobachtet, wenn geschorene anstatt nicht geschorener Hautpräparate punktiert wurden. Der größte Flächendurchmesser und die höchste Anzahl von Hautstanzen je Punktion, wurden mit 18G Kanülen im Vergleich zu 22G und 18GM Kanülen beobachtet ($p < 0{,}05$). Punktionen mit 18GM Kanülen von nicht geschorenen Hautpräparaten erzielten die niedrigste Anzahl von Hautstanzen. Haarfragmente wurden regelmäßig mit Hautstanzen verschleppt. Die Anzahl der verschleppten Haarfragmente war bei Verwendung geschorener Hautpräparate höher als bei nicht geschorenen. Hierbei produzierten 18GM Kanülen die höchste Anzahl an Haarfragmenten, und 22G Kanülen die niedrigste Anzahl bei beiden Präparationsarten.

In den Paraffinschnitten wurden verschiedene dermale und epidermale Strukturen, wie Bindegewebe (14/16 Proben) Stratum corneum (13/16), Epidermis (12/16), isolierte Haarschäfte und Schweißdrüsen (11/16), Haarschäfte eingebettet in Epidermis und Talgdrüsen (6/16) und Haarbälge (5/16), identifiziert.

Hautstanzen können Bakterien enthalten und verschleppen, was möglicherweise zur Entstehung von Infektionen nach Injektionen beitragen kann. Die Inzidenz, Größe und Menge von Hautstanzen, sowie die Menge an verschleppten Bakterien, kann durch die Verwendung möglichst kleiner Kanülen und einer entsprechenden Hautdesinfektion vor der Injektion, reduziert werden. Letztendlich ist davon auszugehen, dass vorbereitende Maßnahmen vor einer Injektion das Risiko einer Keimverschleppung und dadurch bedingten Infektion nicht vollständig ausschließen können.

5 Summary

Tobias Puschmann:

Histologic and microbiologic evaluation of skin cores obtained from skin punctures, depending on needle type and disinfection and preparation methods of the injection site, in horses

The aim of this study was to assess the impact of different skin disinfection and preparation methods on the bacterial flora of the injection site and of skin cores produced through needle punctures. Furthermore, the occurrence of skin cores and hair fragments was investigated, as well as their histologic components.

Skin specimens non-prepared (*no-clip*), disinfected (*dis.no-clip*), clipped and disinfected (*dis.clip*) and clipped and antiseptically washed (*as.clip*) were obtained from euthanized horses. To produce skin cores, the skin specimens were punctured with 18G needles (18G), 18G needles with stylet (18GM), and 22G needles. The needles were flushed with sterile saline solution.
Efficacies of disinfection protocols were compared using swab samples and skin biopsies. For the microbiological evaluation of skin cores, the saline was collected in blood culture systems or homogenization test tubes, containing either phosphated buffered saline solution or culture medium. Samples were processed qualitatively (ql) or semi-quantitatively (sqt).
For microscopic evaluation of skin core incidence, size, and number of skin cores per puncture, the saline was collected on slides and Pappenheim staining was performed. Skin cores, produced with 18G, were also fixed in a 4% formalin solution, after which paraffin embedding and sectioning was performed. Paraffin sections were stained with Masson's trichrome stain. Light microscopy was used for evaluation.

Overall, different bacterial species were isolated from the samples, including equine skin commensals and others, such as fecal contaminants. Comparing disinfection protocols, bacterial isolates were reduced significantly if skin specimens were disinfected or antiseptically prepared ($p < 0.05$), whereas clipping did not produce additional significant decreases. The same tendencies were observed when bacterial isolation of 18G needle contents was compared with the different preparation protocols ($p < 0.05$). The lowest amount of bacteria were isolated from 22G punctures, and the highest amount was detected when 18GM needles were used (ql; *no-clip*; non-significant).

Depending on preparation protocols and needle types, skin cores were detected in many needle punctures. Displaced skin cores were more frequently observed when clipped, instead of unclipped specimens were punctured. The largest surface area and highest number of skin cores was observed with 18G compared to 22G and 18GM ($p < 0.05$). Punctures with 18GM *no-clip* resulted in the lowest number of skin cores. Hair fragments were regularly translocated along with skin cores. Overall more hair fragments were displaced in clipped compared to unclipped samples, with the highest amount detected in 18GM, and the lowest with 22G needles with both preparation types.

Paraffin sections identified several different epidermal and dermal structures, such as connective tissue (14/16 samples), stratum corneum (13/16), epidermis (12/16), isolated hair shafts and sweat glands (11/16), hair shafts integrated with epidermis and sebaceous glands (6/16), and hair bulbs (5/16).

We conclude that skin cores can contain and transfer bacteria, which may contribute to infective post injection complications. However, the bacterial load and the incidence, skin core size and number can be reduced by using smaller needles and performing appropriate disinfection methods prior to injections. Nevertheless, it must be assumed, that the risk of transferring bacteria and causing an infection, will not be completely eliminated by preparation and disinfection methods.

6 Literaturverzeichnis

ADAMS, S. B., G. E. MOORE, M. ELRASHIDY, A. MOHAMED u. P. W. SNYDER (2010):
Effect of needle size and type, reuse of needles, insertion speed, and removal of hair on contamination of joints with tissue debris and hair after arthrocentesis.
Vet. Surg. 39, 667-673

AMMANN, K. (1954):
Die forensische Bedeutung der Injektionsschäden.
Schweiz. Arch. Tierheilkd. 96, 569-579

ANDREW, J. H. (1976):
Pulmonary skin embolism: a case report.
Pathology 8, 185-187

BARIS, S., F. GULDOGUS, Y. S. BARIS, D. KARAKAYA u. E. KELSAKA (2004):
Is tissue coring a real problem after caudal injection in children.
Pediatr. Anesth. 14, 755-758

BELAVY, D. (2010):
Brief reports: regional anesthesia needles can introduce ultrasound gel into tissues.
Anesth. Analg. 111, 811-812

BERNAU, A., C. CORNELIUS, W. DAUBER, G. DIETRICH u. P. HEEG (1985):
Hautstanzzylinder bei Gelenkpunktionen.
Orthop. Praxis 5, 359-365

BRANDUS, V. (1968):
The spinal needle as a carrier of foreign material.
Can. Anaesth. Soc. J. 15, 197-201

CAMPBELL, D. C., J. M. DOUGLAS u. G. TAYLOR (1996):
Incidence of tissue coring with the 25-gauge Quincke and Whitacre spinal needles.
Region. Anesth. Pain. M. 21, 582-585

DEEGEN, E. (2003):
Intramuskuläre Injektionen und Komplikationen.
in: 10. Bpt-Kongress, Münster 2003.
Vortragsband Kleintiere, Pferd, p.100-101

EIKMEIER, H. (1990 a):
Allgemeine Sorgfaltspflichten des Tierarztes.
In: H. EIKMEIER, E. FELLMER und H. MOEGLE (Hrsg.): Lehrbuch der gerichtlichen Tierheilkunde
Verlag Paul Parey, Berlin, Hamburg, S. 85-87

EIKMEIER, H. (1990 b):
Injektionen.
In: H. EIKMEIER, E. FELLMER und H. MOEGLE (Hrsg.): Lehrbuch der gerichtlichen Tierheilkunde
Verlag Paul Parey, Berlin, Hamburg, S. 88-95

GERBER, H., P. TERCIER u. A. MÜLLER (1980):
Injektionstechnik - Injektionsfolgen beim Pferd.
Schweiz. Arch. Tierheilkd. 122, 205-215

GESELLSCHAFT FÜR PFERDEMEDIZIN, Dortmund (2013):
Zur Sorgfalt bei i.v., i.m., und s.c. Injektionen und intravenösen Infusionen beim Pferd
In: Auswahl aktueller Leitfäden zu häufigen tierärztlichen Tätigkeiten in der Pferdepraxis
Gesellschaft für Pferdemedizin, Dortmund, S. 11-12

GIBSON, T. u. W. NORRIS (1958):
Skin fragments removed by injection needles.
Lancet 2, 983-985

GULDOGUS, F., Y. S. BARIS, S. BARIS, D. KARAKAYA u. E. KELSAKA (2008):
Comparing tissue coring potentials of hollow needles without stylet and caudal needles with stylet: an experimental study.
Eur. J. Anaesth. 25, 498-501

HAGUE, B. A., C. M. HONNAS, R. B. SIMPSON u. J. G. PELOSO (1997):
Evaluation of skin bacterial flora before and after aseptic preparation of clipped and nonclipped arthrocentesis sites in horses.
Vet. Surg. 26, 121-125

HAINDL, H. u. O. KLINGE (1989):
Post injection abscess caused by non-suitable cannulas?
Biomed. Eng. 34, 268-271

HAPKE, H.-J. (1981):
Störwirkungen bei Tierarzneimitteln.
Dtsch. tierarztl. Wochenschrift 88, 274-278

INNES, J. R., E. J. DONATI u. P. P. YEVICH (1958):
Pulmonary lesions in mice due to fragments of hair, epidermis and extraneous matter accidentally injected in toxicity experiments.
Am. J. Pathol. 34, 161-167

KAST, A. u. Y. TSUNENARI (1983):
Hair embolism in lungs of rat and rabbit caused by intravenous injection.
Lab. Anim. 17, 203-207

KOLLER, A. (1947):
Injektion und Infektion.
Zürich, Veterinärmed. Fak., Diss.

LITTLE, D. C. (1955):
Skin fragments in endopening needles.
Can. Med. Assoc. J. 72, 374-375

MORTON, D., J. A. SAFRON, J. GLOSSON, D. W. RICE, D. M. WILSON u. R. D. WHITE (1997):
Histologic lesions associated with intravenous infusions of large volumes of isotonic saline solution in rats for 30 days.
Toxicol. Pathol. 25, 390-394

OHNESORGE, B., S. PFALZGRAF, K. ROHN, M. NEUHAUS u. E. DEEGEN (2006):
Unverträglichkeitsreaktionen nach intramuskulärer Injektion beim Pferd - Auswertung einer Tierärztebefragung.
Pferdeheilkd. 22, 337-346

ÖZYURT, G., E. MOĞOL, S. TOLUNAY u. B. KERIMOĞLU (2000):
Tissue coring with spinal needles.
Region. Anesth. Pain. M. 25, 665

PIESOLD, J., W. MÜLLER u. J. DREIßIG (1998):
An experimental study on the aspirating reliability of different types of injection syringes with regard to the formation of punch cylinders.
Brit. J. Oral. Max. Surg. 36, 39-43

PUOLAKKA, R., L. C. ANDERSSON u. P. H. ROSENBERG (2000):
Microscopic analysis of three different spinal needle tips after experimental subarachnoid puncture.
Region. Anesth. Pain. M. 25, 163-169

SCHNEIDER, P. u. G. PAPPRITZ (1976):
Hairs causing pulmonary emboli. A rare complication in long-term intravenous studies in dogs.
Vet. Pathol. 13, 394-400

STANEK, C. (2003):
Zur tiefen Infiltration der langen Rückenmuskulatur beim Pferd: Komplikationen und forensische Aspekte.
Pferdeheilkd. 19, 46-48

TEKELI, S. (1974):
Occurrence of hair-fragment emboli in the pulmonary vascular system of rats.
Vet. Pathol. 11, 482-485

TUNALI, Y., G. KAYA, G. TUNALI, S. SOLAKOGLU, S. YENICE u. M. BAHAR (2006):
Detection of epithelial cell transfer in spinal areas by light microscopy and determining any tissue coring via cell culture during combined spinal-epidural interventions.
Region. Anesth. Pain. M. 31, 539-545

WAGENKNECHT, F. (1924):
Beitrag zur Technik der intravenösen Injektion beim Pferde: mechanische Vorbereitung des Operationsfeldes.
Berlin, tierärztl. Hochsch., Diss.

WAHL, K., S. B. ADAMS u. G. E. MOORE (2012):
Contamination of joints with tissue debris and hair after arthrocentesis: the effect of needle insertion angle, spinal needle gauge, and insertion of spinal needles with and without a stylet.
Vet. Surg. 41, 391-398

WILLENER, A. W. (1948):
Mechanik der Injektion und Verunreinigung des Stichkanals.
Bern, Veterinärmed. Fak., Diss.

ZELLER, R. (1958):
Über die Haftpflicht des Tierarztes bei Injektionen.
Prakt. Tierarzt 39, 10-13

ZUBROD, C. J., K. D. FARNSWORTH u. J. L. OAKS (2004):
Evaluation of Arthrocentesis Site Bacterial Flora before and after 4 Methods of Preparation in Horses with and without Evidence of Skin Contamination.
Vet. Surg. 33, 525-530

7 Danksagung

Mein herzlicher Dank geht an Herr Prof. Dr. Bernhard Ohnesorge, der mir die Durchführung dieser Arbeit ermöglicht hat. Vielen Dank für die vielen Stunden in Ihrem Büro, mit stehts guter Laune, unermüdlichen Erklärungen und Hilfestellungen, schnellen Lösungsansätzen, und das Vertrauen in meine Fähigkeiten und die Bereitschaft mich weiter auszubilden.

Auch Frau Prof. Dr. Pfarrer möchte ich herzlich für die Betreuung des histologischen Arbeitsteils danken. Vielen Dank für die unermüdliche Hilfe bei der Etablierung der Methodik, der Versuchsdurchführung und für die konstruktiven Korrekturen.

Ein weiterer herzlicher Dank geht an Frau Dr. Verspohl, für die Betreuung des mikrobiologischen Arbeitsteils. Vielen Dank für die Unterstützung bei der Etablierung der Methodik, der Versuchsdurchführung und für die konstruktiven Korrekturen.

Ich möchte mich herzlich bei Herr Prof. Dr. Karsten Feige und der Klinik für Pferde für die Möglichkeit diese Arbeit durchführen zu können bedanken.
Ein besonderer Dank geht an Herr Dr. Karl Rohn für die unermüdliche Unterstützung mit der statistischen Auswertung.
Vielen Dank an Doris Walter und Marion Gähle aus der Histologie und an alle Mitarbeiter des Mikrobiologischen Instituts für die Unterstützung bei der Versuchsdurchführung.
Außerdem bedanke ich mich bei allen Interns, Assistenzärzten, Doktoranden der Klinik für Pferde und den Studierenden für die Assistenz bei der Versuchsdurchführung und eine tolle Teamarbeit.
Ich danke meiner lieben Familie für die seelische Unterstützung, die Motivation, den Glauben an meine Fähigkeiten und für die finanzielle Unterstützung. Durch euch ist all das möglich gewesen.
Vielen Dank an Janine Wisweh, Tanja Pudert, Maria Müller, Christian Dancker, Isa Kern, Marianne Boeck, Birte Tietgen, und Bettina Seeger für eure Freundschaft, die tolle Zeit, die Motivation, und die seelische Unterstützung.

Danke Christian Niebuhr für dich und unser Leben.